UWB Theory and Applications

UWB Theory and Applications

Edited by

Ian Oppermann, Matti Hämäläinen and Jari Iinatti
All of CWC, University of Oula, Finland

John Wiley & Sons, Ltd

Other Wiley Editorial Offices

John Wiley & Sons Inc., 111 River Street, Hoboken, NJ 07030, USA

Jossey-Bass, 989 Market Street, San Francisco, CA 94103-1741, USA

Wiley-VCH Verlag GmbH, Boschstr. 12, D-69469 Weinheim, Germany

John Wiley & Sons Australia Ltd, 33 Park Road, Milton, Queensland 4064, Australia

John Wiley & Sons (Asia) Pte Ltd, 2 Clementi Loop #02-01, Jin Xing Distripark, Singapore 129809

John Wiley & Sons Canada Ltd, 22 Worcester Road, Etobicoke, Ontario, Canada M9W 1L1

Wiley also publishes its books in a variety of electronic formats. Some content that appears in print may
not be available in electronic books.

British Library Cataloguing in Publication Data

A catalogue record for this book is available from the British Library

ISBN 0-470-86917-8

Typeset in 10/12pt Times by Integra Software Services Pvt. Ltd, Pondicherry, India
Printed and bound in Great Britain by TJ International Ltd, Padstow, Cornwall
This book is printed on acid-free paper responsibly manufactured from sustainable forestry in which at least
two trees are planted for each one used for paper production.

dedicated to

our families

Contents

Preface

The work covered in this book has been undertaken at the Centre for Wireless Communication (CWC) at the University of Oulu, Finland. The authors have been involved with ultra-wideband (UWB) projects for several years, which have included fundamental studies as well as design–build–test projects. A substantial number of propagation measurements have been undertaken as well as work developing simulators, antenna components and prototypes.

The book focuses very much on impulse radio UWB techniques rather than multiband systems. The reasons for this are both practical and historical. The promise of UWB was low complexity, low power and low cost. Impulse radio, being a baseband technology, holds the most promise to achieve these three benefits. The newer multiband proposals may potentially offer the most spectrally efficient solutions, but they are substantially more complex and it is potentially more difficult to ensure compliance with Federal Communications Commission (FCC) requirements.

The historical reason for the focus on impulse radio techniques is that CWC has been working on UWB devices based on impulse radio techniques since 1999. Much of the work in this book has been performed as part of projects carried out at CWC.

At the time of writing, the European regulatory bodies are still to decide on the spectrum allocation mask for Europe. It is expected to be very similar to the FCC mask, but with more stringent protection for bands below 3.1 GHz. Europe's decision on UWB will have a dramatic impact on the size and shape of the market for UWB devices worldwide. Europe is definitely aware of the historical battles of wireless Local Area Network systems (Hiperlan versus IEEE 802.11) and is seeking a harmonized, global approach to standardization and regulation. The race for UWB consumer devices is moving quickly but is definitely not over yet.

Ian Oppermann
Matti Hämäläinen
Jari Iinatti
Oulu, July 2004

Acknowledgements

The work in this book has predominantly been carried out in projects at CWC in the last several years. The contributing projects include FUBS (future UWB systems), IGLU (indoor geo-location solutions), ULTRAWAVES (UWB audio visual entertainment systems) and URFA (UWB RF ASIC). More information about each of these projects may be found on the CWC WWW site http://www.cwc.oulu.fi/home.

The UWB projects at CWC have been funded by the National Technology Agency of Finland (TEKES), Nokia, Elektrobit, Finnish Defence Forces and European Commission. We are most grateful to the financiers for their interest in the subject.

Many researchers have also contributed to this work. The editors would like to thank Ulrico Celentano, Lassi Hentilä, Taavi Hirvonen, Veikko Hovinen, Pekka Jakkula, Niina Laine, Marja Kosamo, Tommi Matila, Tero Patana, Alberto Rabbachin, Simone Soderi, Raffaello Tesi, Sakari Tiuraniemi and Kegen Yu for their contributions.

The authors also offer a special thanks to Mrs. Therese Oppermann for many hours of proof reading and Ms. Sari Luukkonen for taking care of the proofread corrections.

Abbreviations

A	Gain
A_{peak}	Peak amplitude
a_o	first signal component
a	power scaling constant
$a_n^{(t)}$	amplitude gain for n^{th} multipath component
B	Bandwidth
B_e	Bandwidth expansion factor
B_f	Fractional bandwidth
$c_j^{(k)}$	Pseudo random time-hopping code
C	Capacitance, capacitor
C_{gd}	Gate to drain capacitance
C_{gs}	Gate to source capacitance
C_{int}	Integrating capacitor
C_L	Load capacitance
C_{ox}	Gate capacitance
C_n	Node capacitance
$c(t), C(t)$	Spreading code
C_i	Chip (bit of the spreading code)
C_u	number of cells in uncertainty region
D_{opt}	Optimum number of rake branches
D	number of rake fingers
d	distance
d_o	reference distance
d_j	data bit
E_w	Energy
f	Frequency
Δf	Frequency shift
f_c	Centre frequency
f_0	Nominal centre frequency
f_H	Upper frequency
f_i	Carrier frequency
f_L	Lower frequency
f_{max}	Maximum frequency
f_n	Node frequency
f_{PRF}	Pulse repetition frequency

f_T	Transit frequency
g_i	Chip (bit of the spreading code)
g_m	Transconductance, small signal
G_m	Transconductance, large signal
h_{RX}	receiver antenna height
h_{TX}	transmitter antenna height
$h(t)$	Impulse response
i_{dj}	Small signal drain current
i_{out}	Small signal output current
i_{sj}	Small signal source current
I	Current
I_{bias}	Bias current
$I_{D1}, I_{D2} \dots$	Drain currents of transistor M1, M2...
I_0	The zeroth order modified Bessel function of the first kind
I_{out}	Output current
I_{SS}	Current of tail current source
j	Number of current monocycle
k	User
k	Transconductance parameter
k	k-factor of Ricean faded signal
K	SNR value for Ricean fading channel
L	Number of multipath
L	Length of a MOS transistor, Number of multipath
L_r	Number of rake branches
L_c	number of clusters
M	number of simultaneous users
$M1, M2, \dots$	MOS transistor 1, 2, ...
n	Number of bits, body effect factor, attenuation factor
N	Noise power
NP_{10dB}	Number of paths within 10dB of the peak
$NP\,(85\%)$	Number of paths capturing 85% of the energy
N_{smp}	Number of sample points
NF	Noise figure
$Nfft$	Length of IFFT
N	Number of pulses per data symbol
N_U	Number of users
P_d	probability of detection
P_{fa}	probability of false alarm
P_d^{ov}	overall probability of detection
P_m^{ov}	overall probability of missing a code
PG_4	Processing gain from the pulse repetition
PG_2	Processing gain due to the low duty cycle
PG	Processing gain
$P_{TX,av}$	Average single pulse power
$P_{TX,fr}$	Average power over time hopping frame
PL	Attenuation factor

r	Distance
r_n	Node resistance
r_{s1}	Source resistance of M1
$r(t)$	Transmitted signal
R	Resistance, data rate
R_S	Symbol rate
$s(k)$	Laplace operator, non-centrality parameter of Ricean distribution function
$s(t)$	information signal
S_{tr}	Received waveform
$\mathbf{S}$	Signal power
S_{RX}	Received power spectral density
S_{TX}	Transmitted power spectral density
t	Time
t_{coh}	coherence time
$t^{(k)}$	Clock for user k
t_{tr}	Time of flight
t_{sw}	Sweeping time
ΔT	Time delay used in PPM, modulation index
T_c	Time hopping interval inside a frame (thus, chip length)
T_f	Time hopping frame
T_l^i	Delays of the lth cluster
T_p	Pulse width
T_{PRF}	Pulse repetition interval, length of a time frame
T_s	Symbol time
T_i	time to evaluate a decision variable
T_{acq}	acquisition time
T_{fa}	penalty time
T_h	threshold
T_{MA}	mean acquisition time
T_s	time limit
T_w	time delay between the pulses in doublet
$y(t)$	Received signal
$V(t)$	pulse waveform
$V(f)$	pulse spectrum
V_{in}	Small signal input voltage
V_i^+	Positive single-ended, small signal input voltage
V_i^-	Negative single-ended, small signal input voltage
V	Voltage
V_{bias}	Bias voltage
V_{eff}	Effective gate to source voltage
V_{GS}	Gate to source voltage
V_i	Input voltage
V_{int}	Voltage across integrating capacitor
V_{out}	Output voltage
V_{RF}	RF input voltage
V_T	Threshold voltage of a MOS transistor

V_X	Upper input voltage for Gilbert cell
V_Y	Lower input voltage for Gilbert cell
w_{gi}	i^{th} derivative of the Gaussian pulse
w_{tr}	Transmitted waveform
W	Width of a MOS transistor
X_i	Shadowing effect
α	Exponential decay coefficient
$\alpha_{k,l}^i$	Multipath gain coefficients
β	Exponential decay constant
$\delta d_{\lfloor j/N_s \rfloor}^{(k)}$	Early/late data modulation
δ_{opt}	optimal modulation index
δ	modulation index
γ	Ray decay factor
Γ	Cluster decay factor
λ	Ray arrival rate
Λ	Cluster arrival rate
Ω_0	The mean energy of the first path of the first cluster
σ	Standard deviation
σ_1	standard deviation for cluster lognormal fading
σ_2	standard deviation for ray lognormal fading
σ_{Tacq}	Variance of the acquisition time
σ_x	standard deviation for lognormal shadowing
$\tau_{k,l}^i$	Delays for the kth multipath component
τ_{max}	Maximum detectable delay
$\tau_n^{(t)}$	excess delay
τ_e	code phase
μ_n	Electron mobility near silicon surface
ω_{ti}	Unity-gain frequency of integrator
ξ_l	Fading term
ξ	signal-to-noise ratio
a.c.	Alternating Current
AC	Absolute Combining
ACF	Autocorrelation Function
ACK	Acknowledgement
ADSL	Asymmetric Digital Subscriber Line
ALT PHY	Alternative Physical
AOA	Angle of arrival
A-rake	all rake receiver
ARQ	Automatic Repeat reQuest
AWGN	Additive White Gaussian Noise
BER	Bit Error Rate
BiCMOS	Bipolar Complementary Metal-Oxide-Semiconductor process
BJT	Bipolar Junction Transistor
BPAM	Binary Pulse Amplitude Modulation
CAP	Contention Access Period
CLPDI	chip level post detection integration algorithm

CTA	Channel Time Allocation
CTAP	Channel Time Allocation Period
CTS	clear to send
CDMA	Code Division Multiple Access
CEPT	European Conference of Postal and Telecommunications
CIR	Channel Impulse Response
CMOS	Complementary Metal-Oxide-Semiconductor
CFAR	constant false alarm rate
CMF	Code Matched Filter
CSMA/CA	Carrier sense multiple access with collision avoidance
DAB	Digital Audio Broadcasting
DARPA	Defense Advanced Research Projects Agency, USA
DC	Direct Current
DCOP	Direct Current Operation Point
DEV	Device
DM	Deterministic Model
DME	Device Management Entity
DS	Direct Sequence
DSO	Digital Sampling Oscilloscope
DSSS	Direct Sequence Spread Spectrum
DUT	Device Under Test
DVB	Digital Video Broadcasting
EGC	Equal GainCombining
ETSI	European Telecommunications Standards Institute
FCC	US Federal Communications Commission
FAA	Federal Aviation Administration, USA
FCC	Federal Communications Commission, USA
FD	Frequency Domain
FEM	Finite element method
FDTD	Finite difference time domain
FET	Field Effect Transistor
FFT	Fast Fourier Transform
FIR	Finite Impulse Response,
GaAs	Gallium Arsenide
GaN	Gallium Nitride
GaP	Gallium Phosphide
Gm-C	Transconductor-Capacitor
GPR	Ground Penetrating Radar
GPS	Global Positioning System
GSM	Global System for Mobile Communications
HBT	Hetero-junction Bipolar Transistor
HDR	high data rate
HEMT	High Electron Mobility Transistor
IC	Integrated Circuit
IEEE	The Institute of Electrical and Electronics Engineers
IF	Intermediate Frequency

IFFT	Inverse fast Fourier transform
I/H	Integrate and Hold
InP	Indium Phosphide
InSb	Indium Antimonide
IR	Impulse Radio
IRA	Impulse radiating antenna
I-Rake	Ideal rake receiver
ISI	Inter Symbol Interference
ISM	Industrial, Scientific and Medical
ISO	International Standards Organization
ITU	International Telecommunications Union
LDR	Low data rate
LLC	Logical Link Control
LLNL	Lawrence Livermore National Laboratory
LNA	Low Noise Amplifier
LO	Local Oscillator (frequency)
LOS	Line-of-Sight
LPD	Low Probability of Detection
LPDA	Log-periodig dipole array
LPI	Low Probability of Interception
MAC	Medium Access Control
MBT	Modified Bowtie
MC	Multi-Carrier
MCTA	Management Channel Time Allocations
MESFET	Metal Semiconductor Field Effect Transistor
MF	Matched Filter
ML	Maximum Likelihood
MLME	MAC Layer Management Entity
MMIC	Microwave/Millimetre-wave Integrated Circuit
MoM	Methods of moments
MRC	Maximum Ratio Combining
MT	Multi-tone
NLOS	Non-Line-of-Sight
NMOS	N-channel Metal-Oxide-Semiconductor FET
NOI	Notice of Inquiry
OOK	On-off Keying
OSI	Open System Interconnection
PA	Power Amplifier
PAM	Pulse Amplitude Modulation
PCB	Printed circuit board
PE	Power Estimation
PER	Packet Error Rate
PG	Processing Gain
PHY	Physical layer
PLL	Phase Locked Loop
PMOS	P-channel Metal-Oxide-Semiconductor

PN	Pseudo-random Noise
PNC	PicoNet Coordinator
PPM	Pulse Position Modulation
PR	Pseudo Random
P-rake	Partial rake receiver
PRI	Pulse Repetition Interval
PRF	Pulse Repetition Frequency
PSD	Power Spectral Density
PSM	Pulse Shape Modulation
PTD	Programmable Time Delay
PVT	Process, power supply Voltage and Temperature
PPM	Pulse Position Modulation
QoS	Quality of Service
RC	Resistor-Capacitor (circuit)
RF	Radio Frequency
RFIC	Radio Frequency Integrated Circuit
RMS	Root Mean Square
RTS	Request to send
RX	Receiver, Receiver port
S1, S2	Short Pulses 1, 2
SC	Switched Capacitor
S/H	Sample and Hold (circuit)
SIFS	Short Inter-frame spacing
SiGe	Silicon Germanium semiconductor process
SINR	Signal-to-Interference-plus-Noise Ratio
SIR	Signal-to-Interference Ratio
SM	Statistical Model
SNR	Signal-to-Noise Ratio
S-rake	Selective rake receiver
SRD	Step Recovery Diode
SS	Spread Spectrum
SV	Saleh-Valenzuela Channel Model
TD	Time Domain
TDMA	Time division multiple access
TDOA	Time difference of arrival
TH	Time-Hopping
TH-PPM	Time-Hopping Pulse Position Modulation
TM	Time Modulated
TOA	Time of Arrival
TR	Transistor
TSA	Tapered slot antenna
TX	Transmitter, Transmitter port
UMTS	Universal Mobile Telecommunications System
UMB	Ultra-WideBand
UWBWG	Ultra-Wideband Working Group
VCO	Voltage Controlled Oscillator

VHF	Very high frequency
VNA	Vector Network Analyser
VSWR	Voltage standing wave ratio
WLAN	Wireless Local area network
WO	Worst case One
WP	Worst case Power
WPAN	Wireless Personal Area Networks
WS	Worst case Speed
WZ	Worst case Zero

1

Introduction

Ian Oppermann, Matti Hämäläinen, Jari Iinatti

1.1 Introduction

The world of ultra-wideband (UWB) has changed dramatically in very recent history. In the past 20 years, UWB was used for radar, sensing, military communications and niche applications. A substantial change occurred in February 2002, when the FCC (2002a,b) issued a ruling that UWB could be used for data communications as well as for radar and safety applications. This book will focus almost exclusively on the communications aspects of UWB.

The band allocated to communications is a staggering 7.5 GHz, by far the largest allocation of bandwidth to any commercial terrestrial system. This allocation came hot on the heels of the hotly contested, and very expensive, auctions for third generation spectrum in 2000, which raised more than $100 billion for European governments. The FCC UWB rulings allocated 1500-times the spectrum allocation of a single UMTS (universal mobile telecommunication system) licence, and, worse, the band is free to use.

It was no wonder, therefore, that efforts to bring UWB into the mainstream were greeted with great hostility. First, the enormous bandwidth of the system meant that UWB could potentially offer data rates of the order of Gbps. Second, the bandwidth sat on top of many existing allocations causing concern from those groups with the primary allocations. When the FCC proposed the UWB rulings, they received almost 1000 submissions opposing the proposed UWB rulings.

Fortunately, the FCC UWB rulings went ahead. The concession was, however, that available power levels would be very low. If the entire 7.5 GHz band is optimally utilized, the maximum power available to a transmitter is approximately 0.5 mW. This is a tiny fraction of what is available to users of the 2.45 GHz ISM (Industrial, Scientific and Medical) bands such as the IEEE 802.11 a/b/g standards (the Institute of Electrical and Electronics Engineers). This effectively relegates UWB to indoor, short-range,

UWB Theory and Applications Edited by I. Oppermann, M. Hämäläinen and J. Iinatti
© 2004 John Wiley & Sons, Ltd ISBN: 0-470-86917-8

communications for high data rates, or very low data rates for substantial link distances. Applications such as wireless UWB and personal area networks have been proposed, with hundreds of Mbps to several Gbps and distances of 1 to 10 metres. For ranges of 20 metres or more, the achievable data rates are very low compared with existing wireless local area network (WLAN) systems.

One of the enormous potentials of UWB, however, is the ability to move between the very high data rate, short link distance and the very low data rate, longer link distance applications. The trade-off is facilitated by the physical layer signal structure. The very low transmit power available invariably means multiple, low energy, UWB pulses must be combined to carry 1 bit of information. In principle, trading data rate for link distance can be as simple as increasing the number of pulses used to carry 1 bit. The more pulses per bit, the lower the data rate, and the greater the achievable transmission distance.

1.1.1   Scope of this Book

This book explores the fundamentals of UWB technology with particular emphasis on impulse radio (IR) techniques. The goals of the early parts of the book are to provide the essential aspects of knowledge of UWB technology, especially in communications and in control applications. A literature survey examining books, articles and conference papers presents the basic features of UWB technology and current systems. A patent database search provides a historical perspective on the state-of-art technology.

1.2   UWB Basics

Other terms associated with 'ultra-wideband' include 'impulse', 'short-pulse', 'non-sinusoidal', 'carrierless', 'time domain', 'super wideband', 'fast frequency chirp' and 'mono-pulse' (Taylor, 1995).

Impulse radio communication systems and impulse radars both utilize very short pulses in transmission that results in an ultra-wideband spectrum. For radio applications, this communication method is also classified as a pulse modulation technique because the data modulation is introduced by pulse position modulation (PPM). The UWB signal is noiselike which makes interception and detection quite difficult. Due to the low-power spectral density, UWB signals cause very little interference with existing narrow-band radio systems. Depending on the attitude of national and international regulatory bodies, this should allow licence-free operation of radio systems.

Time-modulated (TM) impulse radio signal is seen as a carrier-less baseband transmission. The absence of carrier frequency is the very fundamental character that differentiates impulse radio and impulse radar transmissions from narrow-band applications and from direct sequence (DS) spread spectrum (SS) multi-carrier (MC) transmissions, which can also be characterised as an (ultra) wideband technique. Fast slewing chirps and exponentially damped sine waves are also possible methods of generating UWB signals.

At the end of the book there is an extensive bibliography of UWB technology in general, and particularly impulse radio and impulse radar systems. Impulse radars, sensors, etc., are touched on in this book, but the main focus is in the communication sector.

1.2.1   Advantages of UWB

UWB has a number of advantages that make it attractive for consumer communications applications. In particular, UWB systems

- have potentially low complexity and low cost;
- have noise-like signal;
- are resistant to severe multipath and jamming;
- have very good time domain resolution allowing for location and tracking applications.

The low complexity and low cost of UWB systems arises from the essentially baseband nature of the signal transmission. Unlike conventional radio systems, the UWB transmitter produces a very short time domain pulse, which is able to propagate without the need for an additional RF (radio frequency) mixing stage. The RF mixing stage takes a baseband signal and 'injects' a carrier frequency or translates the signal to a frequency which has desirable propagation characteristics. The very wideband nature of the UWB signal means it spans frequencies commonly used as carrier frequencies. The signal will propagate well without the need for additional up-conversion and amplification. The reverse process of downconversion is also not required in the UWB receiver. Again, this means the omission of a local oscillator in the receiver, and the removal of associated complex delay and phase tracking loops.

Consequently, TM-UWB systems can be implemented in low cost, low power, integrated circuit processes (Time Domain Corporation, 1998). TM-UWB technique also offers grating lobe mitigation in sparse antenna array systems without weakening of the angular resolution of the array (Anderson *et al.*, 1991). Grating lobes are a significant problem in conventional narrowband antenna arrays.

Due to the low energy density and the pseudo-random (PR) characteristics of the transmitted signal, the UWB signal is noiselike, which makes unintended detection quite difficult. Whilst there is some debate in the literature, it appears that the low power, noise-like, UWB transmissions do not cause significant interference to existing radio systems. The interference phenomenon between impulse radio and existing radio systems is one of the most important topics in current UWB research.

Time-modulation systems offer possibility for high data rates for communication. Hundreds of Mbps have been reported for communication links. It is estimated (Time Domain Corporation, 1998; Kolenchery *et al.*, 1997) that the number of users in an impulse radio communication system is much larger than in conventional systems. The estimation is claimed to be valid for both high- and low-data-rate communications.

Because of the large bandwidth of the transmitted signal, very high multipath resolution is achieved. The large bandwidth offers (and also requires) huge frequency diversity which, together with the discontinuous transmission, makes the TM-UWB signal resistant to severe multipath propagation and jamming/interference. TM-UWB systems offer good LPI and LPD (low probability of interception/detection) properties which make it suitable for secure and military applications.

The very narrow time domain pulses mean that UWB radios are potentially able to offer timing precision much better than GPS (global positioning system) (Time Domain Corporation, 1998) and other radio systems. Together with good material penetration properties, TM-UWB signals offer opportunities for short range radar applications such as rescue and anti-crime operations, as well as in surveying and in the mining industry. One should however understand that UWB does not provide precise targeting and extreme penetration at the same time, but UWB waveforms present a better choice than do conventional radio systems.

1.3 Regulatory Bodies

One of the important issues in UWB communication is the frequency allocation. Some companies in the USA are working towards removing the restrictions from the FCC's regulations for applications utilising UWB technology. These companies have established an Ultra-Wideband Working Group (UWBWG) to negotiate with the FCC. Similar discussion on frequency allocation and radio interference should also emerge in Europe. Currently, there are no dedicated frequency bands for UWB applications in the ETSI (European Telecommunications Standards Institute) or ITU (International Telecommunications Union) recommendations.

1.3.1   UWB Regulation in the USA

Before the FCC's first Report and Order (Federal Communications Commission, 2002a,b), there was significant effort by industrial parties to convince the FCC to release UWB technology under the FCC Part 15 regulation limitations, and to allow licence-free use of UWB products. The FCC Part 15 Rules permit the operation of classes of radio frequency devices without the need for a licence or the need for frequency coordination (47 C.F.R. 15.1). The FCC Part 15 Rules attempt to ensure a low probability of unlicensed devices causing harmful interference to other users of the radio spectrum (47 C.F.R. 15.5). Within the FCC Part 15 Rules, intentional radiators are permitted to operate within a set of limits (47 C.F.R. 15.209) that allow signal emissions in certain frequency bands. They are not permitted to operate in sensitive or safety-related frequency bands, which are designated as restricted bands (47 C.F.R.15.205). UWB devices are intentional radiators under FCC Part 15 Rules.

In 1998, the FCC issued a Notice of Inquiry (NOI) (Federal Communications Commission, 1998). Despite the very low transmission power levels anticipated, proponents of existing systems raised many claims against the use of UWB for civilian communications. Most of the claims related to the anticipated increase of interference level in the restricted frequency bands (e.g. TV broadcast bands and frequency bands reserved for radio astronomy and GPS). The Federal Aviation Administration (FAA) expressed concerned about the interference to aeronautical safety systems. The FAA also raised concerns about the direction finding of UWB transmitters.

The organizations that support UWB technology see large scale possibilities for new innovative products utilizing the technology. The FCC Notice of Inquiry and comments can be found on the Internet (Ultra-Wideband Working group, 1998, 1999, 2004).

When UWB technology was proposed for civilian applications, there were no definitions for the signal. The Defense Advanced Research Projects Agency (DARPA) provided the first definition for UWB signal based on the fractional bandwidth B_f of the signal. The first definition provided that a signal can be classified as an UWB signal if B_f is greater than 0.25. The fractional bandwidth can be determined as (Taylor, 1995).

$$B_f = 2\frac{f_H - f_L}{f_H + f_L} \tag{1.1}$$

where f_L is the lower and f_H is the higher $-3\,dB$ point in a spectrum, respectively.

CURRENT UWB DEFINITION

In February 2002, the FCC issued the FCC UWB rulings that provided the first radiation limitations for UWB, and also permitted the technology commercialization. The final report of the FCC First Report and Order (Federal Communications Commission, 2002a,b) was publicly available during April 2002. The document introduced four different categories for allowed UWB applications, and set the radiation masks for them.

The prevailing definition has decreased the limit of B_f at the minimum of 0.20, defined using the equation above. Also, according to the FCC UWB rulings the signal is recognized as UWB if the signal bandwidth is 500 MHz or more. In the formula above, f_H and f_L are the higher and lower $-10\,dB$ bandwidths, respectively. The radiation limits by FCC are presented in Table 1.1 for indoor and outdoor data communication applications.

1.3.2 UWB Regulations in Europe

At the time of writing, regulatory bodies in Europe are awaiting further technical input on the impact of UWB on existing systems. The European approach is somewhat more cautious than that of the USA, as Europe requires that a new technology must be shown to cause little or no harm to existing technologies. The European organizations have, of course, been heavily influenced by the FCC's decision. Currently in Europe, the recommendations for short-range devices belong to the CEPT (European Conference of Postal and Telecommunications) working group CEPT/ ERC/ REC 70- 03 (Ultra-Wideband Working Group, 1999). Generally, it is expected that ETSI/CEPT will follow

Table 1.1 FCC radiation limits for indoor and outdoor communication applications

Frequency in MHz	Indoor EIRP in dBm	Outdoor EIRP in dBm
960–1610	−75.3	−75.3
1610–1990	−53.3	−63.3
1990–3100	−51.3	−61.3
3100–10600	−41.3	−41.3
Above 10600	−51.3	−61.3

the FCC's recommendations but will not necessarily directly adopt the FCC's regulations. The ITU limits (ITU 2002) for indoor and outdoor applications are defined by the formulas represented in Table 1.2.

Figure 1.1 shows the current proposal for the European spectral mask limits as well as the FCC masks. The upper plot represents the masks for data communication applications for indoor and outdoor use. The lower plot gives the FCC radiation mask for radar and sensing applications. In all cases the maximum average power spectral density

Table 1.2 ITU radiation limits for UWB indoor and outdoor applications

	Frequency range [GHz]		
	$f < 3.1$	$3.1 < f < 10.6$	$f > 10.6$
Indoor mask	$-51.3 + 87\log(f/3.1)$	-41.3	$-51.3 + 87\log(10.6/f)$
Outdoor mask	$-61.3 + 87\log(f/3.1)$	-41.3	$-61.3 + 87\log(10.6/f)$

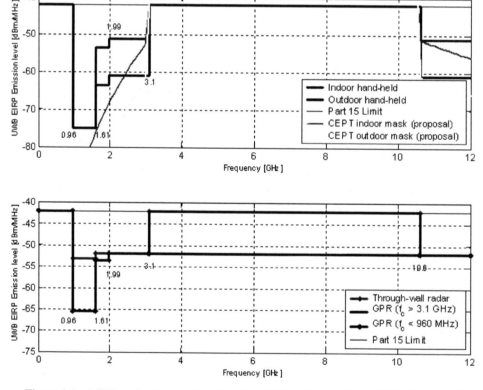

Figure 1.1 UWB radiation mask defined by FCC and the existing CEPT proposal

follows the limit of FCC Part 15 regulations (Federal Communications Commission, 2004).

The working groups for UWB include ERM/TG31A covering generic UWB, and ERM/TG31B, which covers UWB for automotive applications at higher bands.

1.3.2.1   IEEE 802.15.3a

The IEEE established the 802.15.3a study group to define a new physical layer concept for short range, high-data-rate, applications. This ALTernative PHYsical (ALT PHY) is intended to serve the needs of groups wishing to deploy high-data-rate applications. With a minimum data rate of 110 Mbps at 10 m, this study group intends to develop a standard to address such applications as video or multimedia links, or cable replacement. The study group has been the focus of significant attention recently, as the debate over competing UWB physical layer technologies has raged. The work of the study group also includes analysing the radio channel model proposal to be used in the UWB system evaluation.

The purpose of the study group is to provide a higher speed PHY for the existing approved 802.15.3 standard for applications which involve imaging and multimedia (IEEE, 2004). The main desired characteristics of the alternative PHY are:

- coexistence with all existing IEEE 802 physical layer standards;
- target data rate in excess of 100 Mbits/s for consumer applications;
- robust multipath performance;
- location awareness;
- use of additional unlicensed spectrum for high rate WPANs (wireless personal area network).

1.3.2.2   IEEE 802.15.4a

The IEEE established the 802.15.4a study group to define a new physical layer concept for low data rate applications utilizing UWB technology at the air interface. The study group addresses new applications that require only moderate data throughput, but long battery life such as low-rate wireless personal area networks, sensors and small networks.

1.4   Conclusions

The fact that UWB technology has been around for so many years and has been used for a wide variety of applications is strong evidence of the viability and flexibility of the technology. The simple transmit and receiver structures that are possible make this a potentially powerful technology for low-complexity, low-cost, communications. As will be discussed in later chapters, the physical characteristics of the signal also support location and tracking capabilities of UWB much more readily than do existing narrower band technologies.

The severe restrictions on transmit power (less than 0.5 mW maximum power) have substantially limited the range of applications of UWB to short distance–high data rate or low data rate–longer distance applications. The great potential of UWB is to allow flexible transition between these two extremes without the need for substantial modifications to the transceiver.

Whilst UWB is still the subject of significant debate, there is no doubt that the technology is capable of achieving very high data rates and is a viable alternative to existing technology for WPAN; short-range, high-data-rate communications; multimedia applications, and cable replacement. Much of the current debate centres around which PHY layer(s) to adopt, development of a standard, and issues of coexistence and interference.

2

UWB Channel Models

Matti Hämäläinen, Veikkò Hovinen, Lassi Hentilä

2.1 Introduction

In many respects, the UWB channel is very similar to a wideband channel as may be experienced in spread spectrum or CDMA systems. The main distinguishing feature of an 'ultra' wideband channel model is the extremely multipath-rich channel profile. With a bandwidth of several GHz, the corresponding time resolution of the channel is of the order of fractions of a nanosecond. When translated to the spatial domain, this means that it is possible to distinguish reflecting surfaces separated by mere centimetres. Many everyday objects now can be seen to act as distinct specular reflectors rather than contributing to 'lumped' multipath components. The significantly greater time resolution is accompanied by substantially lower power per multipath component.

This chapter examines common UWB channel models, provides methods to measure UWB channels, and introduces the channel model adopted by the IEEE 802.15.3a study group, which will be used as a reference model in UWB system performance studies.

2.2 Channel Measurement Techniques

There are two possible domains for performing the channel sounding to measure the UWB radio channel. First, the channel can be measured in the frequency domain (FD) using a frequency sweeping technique. With FD sounders, a wide frequency band is swept using a set of narrow-band signals, and the channel frequency response is recorded using a vector network analyser (VNA). This corresponds to S_{21}-parameter measurement set-up, where the device under test (DUT) is a radio channel.

Second, the channel can be measured in the time domain (TD) using channel sounders that are based on impulse transmission or direct sequence spread spectrum signalling.

UWB Theory and Applications Edited by I. Oppermann, M. Hämäläinen and J. Iinatti
© 2004 John Wiley & Sons, Ltd ISBN: 0-470-86917-8

With impulse based TD sounders, a narrow pulse is sent to the channel and the channel impulse response is measured using a digital sampling oscilloscope (DSO).

The corresponding train of impulses can also be generated using a conventional direct sequence spread spectrum (DSSS) based measurement system with a correlation receiver. The performance of the DSSS sounder is based on the properties of the auto-correlation function (ACF) of the spreading code used as an overlay signal. The drawback of using the DSSS technique is that it needs very high chip rates to achieve bandwidths required for UWB.

In this chapter, the frequency and time domain measurement concepts are presented. Theoretically, both techniques give the same result if there is a static measurement environment and an unlimited bandwidth.

2.2.1 Frequency Domain Channel Sounding

With frequency domain sounders, the RF signal is generated and received using a vector network analyser (VNA) which makes the measurement set-up quite simple. The sounding signal is a set of narrow-band sinusoids that are swept across the band of interest. The frequency domain approach makes it possible to use wideband antennas, instead of special impulse radiating antennas. As will be discussed in a later chapter, UWB antennas have restrictions, for example, with ringing leading to pulse shape distortion. The UWB channel models can then be generated at the data post-processing stage. When the FD sounder approach is used, the channel state during the soundings must be static to maintain the channel conditions during the sweep. The maximum sweep time is limited by the channel coherence time. If the sweep time is longer than the channel coherence time, the channel may change during the sweep. For fast changing channels, other sounding techniques are needed.

The performance of the frequency domain sounder is also limited by the maximum channel delay. The upper bound for the detectable delay τ_{max} can be defined by the number of frequency points used per sweep and the bandwidth B (frequency span to be swept). This is given by

$$\tau_{max} = (N_{smp} - 1)/B \tag{2.1}$$

where N_{smp} is the number of frequency points.

Another possible source of error in the measurement process is the frequency shift caused by the propagation delay when long cables are used, or when the flight time of the sounding signal is long. In frequency-sweep mode, the sounding signal is rapidly swept across the whole band of interest. For a transmitter and receiver that are in lock-step sweeping across the frequency band of interest, very long propagation delays can cause the receiver to take samples at a frequency that is higher than the received frequency. This frequency shift Δf is a function of the propagation time t_{tr} (time of flight), the frequency span B and the sweep time t_{sw} as

$$\Delta f = t_{tr} \left(\frac{B}{t_{sw}} \right). \tag{2.2}$$

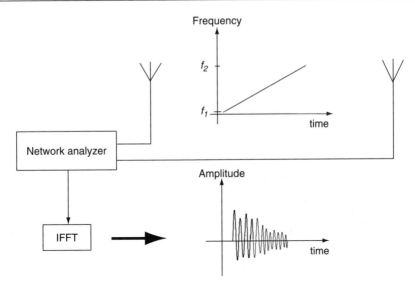

Figure 2.1 Vector network analyser-based frequency domain channel sounding system

In general, Δf has to be smaller than the analyser IF bandwidth to obtain reliable results. The idea used in FD measurements is presented in Figure 2.1. After the channel frequency response has been measured, the time domain representation (impulse response) can be achieved by inverse Fourier transform (IFFT).

2.2.1.1   Signal Analysis Using IFFT

The signal measured using a VNA is a frequency response of the channel. The inverse Fourier transform is used to transform the measured frequency domain data to the time domain. The IFFT is usually taken directly from the measured raw data vector. This processing is possible since the receiver has a down-conversion stage with a mixer device. This method is referred to as the complex baseband IFFT, and is sufficient for modeling narrow- and wideband systems.

There are two common techniques for converting the signal to the time domain, which both lead to approximately the same results. The first approach is based on Hermitean signal processing, which results in a better pulse shape. The second approach is the conjugate approach. Tests show that the conjugate approach is an easier and more efficient way of obtaining approximately the same pulse shape accuracy. These two approaches are introduced next.

2.2.1.2   Hermitian Signal Processing

Using Hermitian processing, the pass-band signal is obtained with zero padding from the lowest frequency down to DC (direct current), taking the conjugate of the signal, and reflecting it to the negative frequencies. The result is then transformed to the time domain using IFFT. This Hermitian method is shown in Figure 2.2. The signal

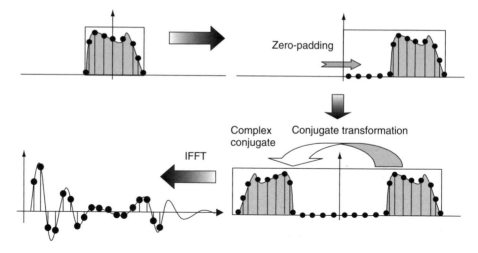

Figure 2.2   Zero padding, conjugate reflection and resulting impulse response

spectrum is now symmetric around DC. The resulting doubled-sided spectrum corresponds to a real signal. The time resolution of the received signal is more than twice that achieved using the baseband approach. This improvement in accuracy is important, since one purpose in UWB channel modelling is to separate accurately the different signals paths.

2.2.1.3   Conjugate Approach

The conjugate method involves taking the conjugate reflection of the passband signal without zero padding. Using only the left side of the spectrum, the signal is converted using the IFFT with the same window size as the Hermitian method. The technique is presented in Figure 2.3. The conjugate result is very likely to be the same as the

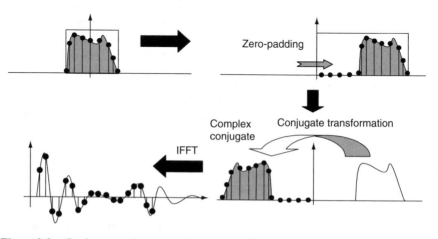

Figure 2.3   Conjugate reflection with zero padding and resulting impulse response

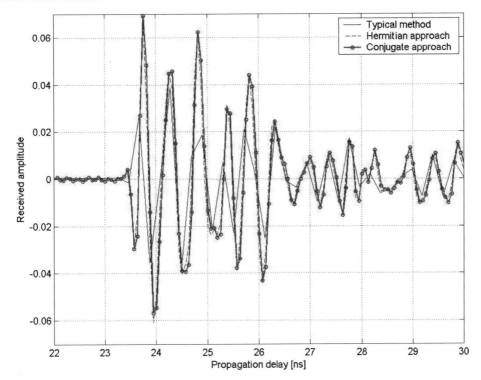

Figure 2.4   Impulse responses of the different IFFT methods

Hermitian result with zero padding. However, the conjugate method is more efficient in terms of data processing complexity, since the matrix calculations in the post-processing stage become easier to manipulate due to the smaller memory requirements.

The impulse responses of these two methods and the baseband method are shown in Figure 2.4, where Hamming window is used in data processing.

From Figure 2.4 it can be seen that for both methods, the main reflection positions are the same and the amplitudes are very close. Thus, the approach based on the left-side conjugate produces an adequate pulse shape with lower processing complexity.

2.2.2   Calibration and Verification

The vector network analyser system, like all measurement systems, requires calibration with the same cables, adapters and other components that will be used for the measurements before the soundings. An enhanced response calibration is required to be able to determine both the magnitude and phase of the transmitted signal (Balanis, 1997). Amplifiers must be excluded from calibration because they are isolated in the reverse direction. The amplifiers' frequency response can be measured independently and their effects can be taken into account in the data post-processing. Long cables and the adapters connected to the ports of the analyser cause a frequency dependent variation in the sounding signal. The deviation is directly proportional to the quality of the used equipments. This variation can, however, be compensated in the calibration procedure.

The calibration process moves the time reference points from the analyser ports to the calibration points at the ends of the cables. When the time references are at the cable ends (at the antenna connectors), the resulting delay profiles only includes the propagation delays that result from the radio channel. Due to their dimensions, delays due to the antennas themselves are insignificant.

The performance of the measurement system was verified in a corridor where major delays could be readily calculated from the room geometry (Hovinen *et al.*, 2002). The physical dimensions of the reference corridor are presented in Figure 2.5. The corresponding impulse response that is calculated from the recorded frequency response is presented in Figure 2.6. Reflections coming from points END I, END II and END III of the corridor (marked on Figure 2.5), as well as the main reflections from the walls, the

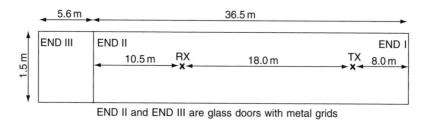

Figure 2.5 Dimensions for the calibration test measurement made in a corridor (figure is not to scale)

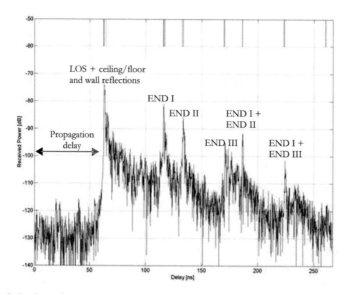

Figure 2.6 Impulse response measured at the corridor presented in Figure 2.5

floor and from the ceiling, can be found with simple calculations. END II and END III are glass doors with a metal grid inside. Comparing the estimated propagation delays plotted at the top of Figure 2.6 with the measured impulse response, a clear conformity between the results can be observed.

2.2.3   Measurement Experimental Set-up

The following section introduces the radio channel measurement system, which consists of a vector network analyser, a wideband amplifier, a pair of antennas and a control computer with LabVIEW™ controlling software. This particular installation has been used in the measurements carried out by CWC.

The network analyser is operating in a response measurement mode, where PORT1 is a transmitter port (TX) and PORT2 is a receiver port (RX). An external amplifier is connected to PORT1 to increase the transmitted power level. The antennas used in the measurement system are CMA-118/A conical antennas developed by Antenna Research Associates, Inc. Typical features of conical antennas are an approximately omnidirectional radiation pattern with a constant phase centre. Both of these features are important in radio channel sounding. The constant phase centre means that the exact radiation point is independent of the frequency. The omnidirectional radiation pattern makes it possible to receive all of the reflections generated by the channel. A low noise amplifier at the input of the VNA can be used to improve the receiver's noise figure, and to increase the energy of the received channel probing signal.

The sweep time is automatically adjusted by the analyser, depending on the bandwidth and on the number of measured frequency points within the sweeping band. Figure 2.7 presents the block diagram of the VNA based measurement set-up. The operating measurement system is presented in Figure 2.8. Figure 2.9 gives examples of the environments where the measurements have been carried out. The left-hand figure introduces the assembly hall at the main building at the University of Oulu, and the right-hand one represents a typical classroom with furniture.

The operation of the measurement set-up, and the applicability of IFFT, was verified by recording the channel frequency response using short cables. The verifications also show that the measurements using short cables (about 1 m) and long cables (maximum 30 m) give the same result for the channel impulse response which proves that the calibration procedure is correct. UWB measurements with high delay resolution generally demonstrate similar structure to wideband measurements for indoor environments. In Figure 2.10, the measured frequency response and the corresponding impulse response are presented. The measurement was performed with one obstructing wall between transmitter and receiver.

2.2.3.1   Modified Frequency Domain Sounding System

The VNA-based measurement system introduced in Figure 2.7 is limited in range by the length of the antenna cables. Due to the fact that the same device works both as a transmitter and receiver, the distance between the antennas is limited to short or

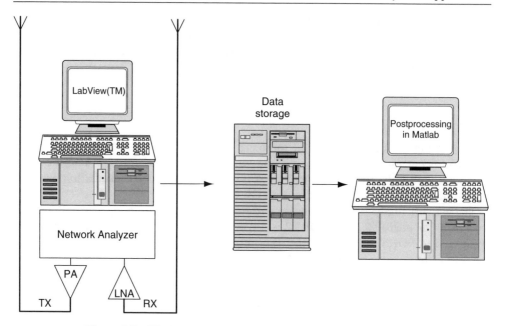

Figure 2.7 The VNA based radio channel measurement set-up

medium range. The measurement set-up can be improved to make it possible to increase the link distance and to make the system more flexible, by changing the topology whilst maintaining the remaining VNA as a receiver (Hämäläinen et al. 2003).

The modified UWB radio channel sounder based on the VNA is presented in Figure 2.11. The sinusoidal probing signal is now generated by an external signal generator. The wired connection between the transmitting and receiving ends is avoided by sending only the triggering signal to the transmitting part via radio. The channel frequency response is again measured using a set of tones much like the approach presented in Section 2.2.1. The functionality of the modified sounding system is presented in Figure 2.11. All of the measurement procedures are again controlled by LabView^{TM} software.

The main component in the modified channel measurement system is a vector network analyser. This device receives the probing signal and makes the calculations for the channel S_{21} values in the same way as described in Section 2.2.1. Rather than coming from the VNA itself, the probing signal is now sent by the external sweeping signal generator, which is SMIQ in the reference system. These two devices are frequency synchronized by using the external reference clock. The transmitting port must be terminated with $50\,\Omega$ impedance to avoid unwanted reflections from the unused RF-port of the VNA.

A low noise amplifier (LNA) can be used at the receiver port of the VNA to improve the noise figure of the receiver, and of course, to amplify the received probing signal level. A power amplifier (PA) can be used at the transmitter end. Any linear power amplifier can be used to have the required power level at the receiver input.

In the test setup, a Rohde & Schwartz AFGU function generator is used to generate the triggering pulses that are fed to the SMIQ and VNA. If the propagation delay

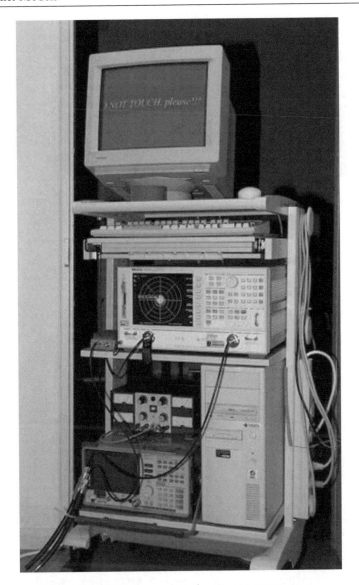

Figure 2.8   Frequency domain measurement system

between the devices is long, the trigger pulse timing for SMIQ and VNA must be tuned to make the SMIQ probing signal and the VNA reference signal arrive at the VNA detector simultaneously. However, the maximum detectable delay τ_{max} cannot be exceeded. The devices are operating in single-sweep stepping mode, sweeping the pre-defined frequency band using 1601 points. Again, the frequency difference needs to be less than the VNA IF-bandwidth for reliable detection.

In the reference system settings, the sweep time is approximately 5 s, and the sweep time depends only on the number of measured frequency points. The number of

Figure 2.9 Different environments covered during the channel measurement campaign

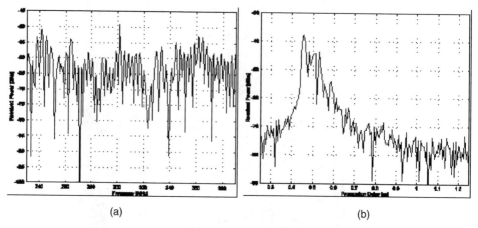

(a)   (b)

Figure 2.10 Example of an ultra-wideband through wall radio channel measurement: (a) frequency response and (b) corresponding impulse response

frequency steps and therefore the sweeping time can be decreased if the maximum delay of the channel is small.

LabView delivers almost all of the timing information and all the commands and adjustments for the devices. A constant clock reference is used to maintain high frequency stability in the system. An external 5 MHz clock reference, based on the TV stripe frequency signal, is fed to the external clock reference inputs in the VNA and the sweeping signal generator. After this procedure, a maximum frequency stability of 10^{-12} is guaranteed. During the frequency sweeps, the phase difference between the generator signal and the analyser signal is not known. However in UWB radio-channel models, the phase information does not play as significant role as it does in wideband models because there is no carrier involved. The carrier would normally provide the reference phase. A scalar network analyser or spectrum analyser can also be used because phase information is not utilized in data post-processing. If the objective is to use a continuous sweep during the measurements, the sweep time in this construction increases significantly, to approximately 30 s.

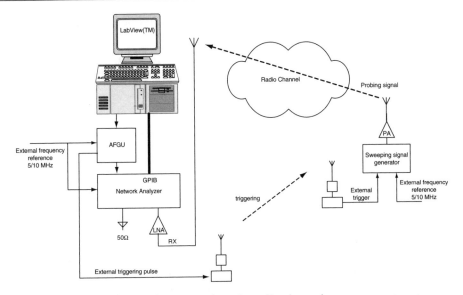

Figure 2.11 Modified frequency domain radio channel measurement system

In the test system, LV217 VHF combat net radios are used to transfer the triggering signal from AFGU to SMIQ. The output pulse of AFGU is transformed to sound wave and sent via LV217. At the SMIQ-end, the sound wave is converted back to an electrical signal and used as an external trigger for the SMIQ. The use of VHF (very high frequency) radios in a triggering link is justified by their frequency band that is outside of the band under the measurements.

The frequency band that can be measured depends on the frequency range of the VNA and SMIQ. The current devices used in the experiment allow the measurements in a selected band between 50 MHz and 20 GHz. The antennas may be selected by the frequency range requirements. However, the antennas should have a constant phase centre and they should be omnidirectional if used in the radio channel measurements in order to minimize distortion caused to the probing signal.

The limitation of the frequency domain channel sounding is that there needs to be a static environment during the recordings.

2.2.4   Time Domain Channel Sounding

As discussed in Section 2.2.1, the frequency domain channel sounder excludes the measurements of the non-stationary channel. However, movement can be supported to a certain extent if the soundings are made in time domain. The following section introduces time domain sounding systems that can be utilized for UWB.

2.2.4.1   Impulse Sounding

One way to carry out time-domain UWB radio channel soundings is to use of very short impulses. The receiver in this case is a digital sampling oscilloscope. The bandwidth of

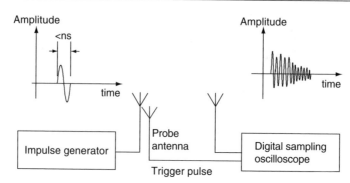

Figure 2.12 Impulse sounder

the sounder depends on the pulse shape and the pulse width used. By changing the pulse width, the spectral allocation can also be changed. However, the simpler the pulse shape, the easier it is to perform the deconvolution during the post-processing, where the channel impulse response is calculated by removing the transmitted pulse waveform from the results.

The most exact channel model can be generated if the pulse waveforms used in the sounding correspond to the waveforms of the application. The topology for the time domain measurements is presented in Figure 2.12.

The impulse-based measurement system requires an additional antenna to be used for triggering purposes. The sounding distance is still limited due to the probing antenna, which is close to the TX antenna needing to be connected to the DSO. However, the long cable is only needed to transfer the triggering pulse for the sounding pulse, so the high quality (e.g., expensive) cable is not needed. In the modified version, the triggering pulse can also be sent via radio link.

2.2.4.2 Direct Sequence Spread Spectrum Sounding

The original wideband radio channel sounding system is based on direct sequence spread spectrum technique and a correlation receiver. Theoretically, a train of impulses can be generated using the maximum length code (m-sequence) and calculating its autocorrelation function at the receiver. The idea of the use of spreading code as a channel probing signal is presented in Figure 2.13(a) where $s(t)$, $r(t)$, $c(t)$ and $h(t)$ are transmitted signal, received signal, pseudo-random code and the channel impulse response, respectively. m-sequence is used to spread the transmitted carrier signal energy over the wide frequency band. The bandwidth of the sounding signal is twice the chip rate, which is the bit rate of the m-sequence.

The well known m-sequence is the optimal pseudo-random code for channel sounding due to the low side-lobe level of its even autocorrelation function (see Figure 2.14). If the length of the m-sequence is N, the normalized even and periodic autocorrelation function has a value of 1 if codes are synchronized, and has a value of $-1/N$ in all the other code phases. The dynamic range of the signal increases as a function of N.

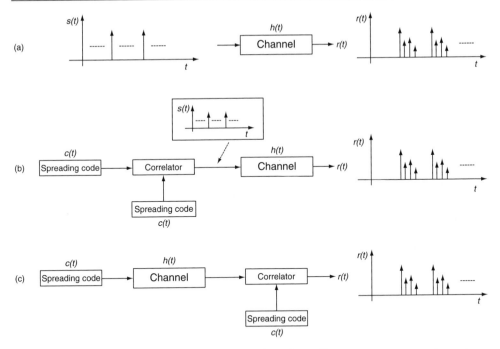

Figure 2.13 Radio channel sounding system based on direct sequence spread spectrum technique

The correlation operation can be performed after the channel [see Figure 2.13(c)] as the radio channel is linear, e.g., the channel does not saturate the propagating signal or create new frequencies.

The correlator output is sampled at the chip rate, and each value represents a certain delay caused by the channel. The magnitude of the sample is related to the strength of the corresponding propagation path. All of this creates a delay resolution which is inversely proportional to the chip rate. The maximum unaliased delay using this sounding method is NT_c, where T_c is the chip length.

In principle, this channel sounding technique can also be applied in the UWB context. Because the bandwidth of the sounding signal depends on the used chip rate, the minimum 500-MHz bandwidth can be achieved using a chip rate of 250 MHz. If the bandwidths are larger than 1 GHz, the chip rate needs to exceed 500 MHz. This high chip-rate requirement limits the utilization of this DSSS in UWB channel sounding. However, if the frequency band to be sounded is located in the lower UWB band, which is allocated for several radar applications, the fractional bandwidth requirements can easily be met.

2.3 UWB Radio Channel Models

Due to the very large bandwidth of the impulse radio signal, the propagation phenomena are different in the lower band and the upper band of the signal spectrum. The size of

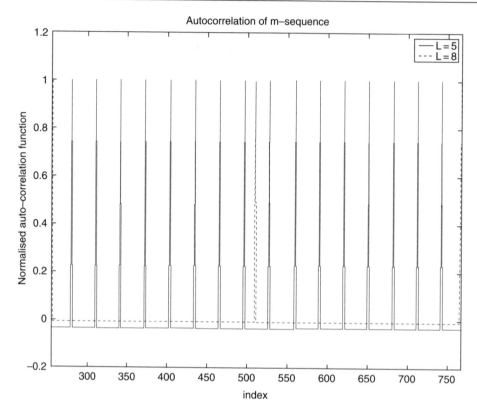

Figure 2.14   Even autocorrelation function of m-sequence

the Fresnel zone, which is a function of wavelength, will be markedly different at the lower and higher frequency ends. The larger Fresnel zone size of the lower frequencies will mean the lower end of the signals will become obstructed more easily, and the signal components will attenuate more than the higher signal components. Also, the natural and man-made interference sources are different in different parts of the spectrum.

For example, the impulse-transmission-based time-domain experimental results show that the radio channel impulse response for a sub-nanosecond pulse is a few hundred nanoseconds long in a typical American laboratory/office building such as that presented by (Win, 1998). The path-loss factor, according to the same measurements, was seen to be between −2.2 and −3.3, which means that the signal power attenuates 22 to 33 dB/decade when the distance increases. According to the same experimental study, the number of resolvable multipaths for ∼ 1 GHz UWB signal bandwidth in an office building is between 5 and 50.

The measurement system used in the indoor measurements (Win, 1998) is also used outdoors to find out UWB's suitability to narrowband vegetation loss models. (Win *et al.*, 1997a) give a comparison between the results from narrowband vegetation loss models (Weissoerger, 1992) and the results based on UWB channel soundings. The results show that the narrowband loss models can be applied to the UWB case.

2.3.1 Modified Saleh–Valenzuela Model

During 2002 and 2003, the IEEE 802.15.3 Working Group for Wireless Personal Area Networks, and especially its channel modelling subcommittee decided to use the so called *modified Saleh–Valenzuela model* (SV) as a reference UWB channel model (Foerster *et al.*, 2003; Foerster, 2003).

The real valued model is based on the empirical measurements originally carried out in indoor environments in 1987 (Saleh and Valenzuela, 1987). Due to the clustering phenomena observed at the measured UWB indoor channel data, the model proposed by IEEE 802.15 is derived from Saleh and Valenzuela using a log-normal distribution rather than an original Rayleigh distribution for the multipath gain magnitude. An independent fading mechanism is assumed for each cluster as for each ray within the cluster. In the SV models, both the cluster and ray arrival times are modelled independently by Poisson processes. The phase of the channel impulse response can be either 0 or π. Therefore the model contains no imaginary component.

The following analytical representation was originally presented by Foerster (2003), who introduces the model characterization in more detail, and also gives the Matlab functions to create the channel realizations. According to Foerster, a discrete time multipath channel impulse can be presented as

$$h_i(t) = X_i \sum_{l=0}^{L_C^{-1}} \sum_{k=0}^{K_{LC}^{-1}} \alpha_{k,l}^i \delta\left(t - T_l^i - \tau_{k,l}^i\right) \tag{2.3}$$

where $\alpha_{k,l}^i$ represents the multipath gain coefficients, T_l^i the delays of the lth cluster, $\tau_{k,l}^i$ gives the delays for the kth multipath component relative to the lth cluster arrival time (T_l^i). Shadowing effect is log-normal distributed and is represented by X_i and i refers to the ith realization.

The proposed modified Saleh-Valenzuela model uses the following definitions:

T_l = the arrival time of the first path of the lth cluster;
$\tau_{k,l}$ = the delay of the kth path within the lth cluster relative to the first path arrival time, T_l;
Λ = cluster arrival rate;
λ = ray arrival rate, i.e. the arrival rate of path within each cluster.

The definition assumes that $\tau_{0,l} = 0$. The cluster arrival time distribution can be presented by (Foerster, 2003)

$$p(T_l|T_{l-1}) = \Lambda \exp[-\Lambda(T_l - T_{l-1})], \quad l > 0 \tag{2.4}$$

and the ray arrival times by

$$p\left(\tau_{k,l}|\tau_{(k-1),l}\right) = \lambda \exp\left[-\lambda\left(\tau_{k,l} - \tau_{(k-1),l}\right)\right], \quad k > 0. \tag{2.5}$$

The channel coefficients are defined by (Foerster, 2003)

$$\alpha_{k,l} = p_{k,l}\xi_l\beta_{k,l}, \text{and} \tag{2.6}$$

$$20\log 10(\xi_l\beta_{k,l}) \propto N(\mu_{k,l}, \sigma_1^2 + \sigma_2^2), \text{or} \tag{2.7}$$

$$|\xi_l\beta_{k,l}| = 10^{(\mu_{k,l}+n_1+n_2)/20}, \tag{2.8}$$

where $n_1 \propto N(0, \sigma_1^2)$ and $n_2 \propto N(0, \sigma_2^2)$ are independent and correspond to the fading on each cluster and ray, respectively, and

$$E\left[|\xi_l\beta_{k,l}|^2\right] = \Omega_0 e^{-T_l/\Gamma}e^{-\tau_{k,l}/\gamma}, \tag{2.9}$$

where Ω_0 is the mean energy of the first path of the first cluster, and $p_{k,l}$ is equiprobable $\{-1, +1\}$ to account for the signal polarity inversion due to the reflections. The $\mu_{k,l}$ is given by (Foerster, 2003)

$$\mu_{k,l} = \frac{10\ln(\Omega_0) - 10T_l/\Gamma - 10\tau_{k,l}/\gamma}{\ln(10)} - \frac{(\sigma_1^2 + \sigma_2^2)\ln(10)}{20}. \tag{2.10}$$

The variables in the above equations represent the fading associated with the lth cluster, ξ_l, and the fading associated with the kth ray of the lth cluster, $\beta_{k,l}$. As noted above, the impulse responses given by the model are real valued.

Due to the fact that the log-normal shadowing of the total multipath energy is captured by the term X_i, the total energy contained in the terms $\{\alpha_{k,l}^i\}$ is normalized to unity for each realization. The shadowing term is characterized by (Foerster, 2003)

$$20\log 10(X_i) \propto N(0, \sigma_x^2). \tag{2.11}$$

The model derives the following channel parameters as an output:

- mean and root mean square (RMS) excess delays;
- number of multipath components;
- power decay profile.

In addition to cluster and ray arrival rates, Λ and λ the model inputs cluster and ray decay factors Γ and γ, respectively, and standard deviation terms in dB for cluster lognormal fading, ray lognormal fading and lognormal shadowing term for total multipath realization σ_1, σ_2 and σ_x, respectively.

Four different channel implementations are suggested, which are based on the average distance between transmitter and receiver, and whether a LOS component is present or not.

Table 2.1 The different SV-models and their main parameters like presented in the IEEE 802.15.3 proposal

Target channel characteristics	SV-1	SV-2	SV-3	SV-4
Mean excess delay (nsec) (τ_m)	5.05	10.38	14.18	
RMS delay (nsec) (τ_{rms})	5.28	8.03	14.28	25
NP_{10dB}			35	
NP (85%)	24	36.1	61.54	
Model parameters				
$\Lambda(1/\text{nsec})$	0.0233	0.4	0.0667	0.0667
$\lambda(1/\text{nsec})$	2.5	0.5	2.1	2.1
Γ	7.1	5.5	14.00	24.00
γ	4.3	6.7	7.9	12
$\sigma_1(\text{dB})$	3.3941	3.3941	3.3941	3.3941
$\sigma_2(\text{dB})$	3.3941	3.3941	3.3941	3.3941
$\sigma_x(\text{dB})$	3	3	3	3
Model characteristics				
Mean excess delay (nsec) (τ_m)	5.0	9.9	15.9	30.1
RMS delay (nsec) (τ_{rms})	5	8	15	25
NP_{10dB}	12.5	15.3	24.9	41.2
NP (85%)	20.8	33.9	64.7	123.3
Channel energy mean (dB)	−0.4	−0.5	0.0	0.3
Channel energy standard (dB)	2.9	3.1	3.1	2.7

SV-1: line-of-sight (LOS) model for 0–4-m
SV-2: Non-LOS (NLOS) model for 0–4-m
SV-3: NLOS for 4–10-m model
SV-4: NLOS for 4–10-m model. This model represents an extreme NLOS multipath channel condition.
NP_{10dB}: Number of paths within 10dB of the peak
NP(85%): Number of paths capturing 85% of the energy

The four channel models and their parameters are listed in Table 2.1 (Foerster, 2003). Figure 2.15 gives an example of 100 channel realizations that are based on SV-3 model.

Figure 2.16 illustrates the difference between the different SV-models in the delay domain. The average profiles are calculated from 100 independent channel realizations, which is the approach recommended by the IEEE 802.15. The delay resolution in the models is 167 ps, which corresponds to a spatial resolution of 5 cm. In Figure 2.17, the number of the distinguishable propagation paths inside a 10-dB dynamic range if compared to the highest path is presented for each SV-model calculated from 100 channel realizations. When the link occupation increases, the number of distinguishable paths also increases. This can easily be seen from the figure.

2.3.2 Other Multipath Models

Radio channel experiments have been carried out at the University of Oulu using the measurement system represented in Figures 2.7 and 2.8. The measured frequency band

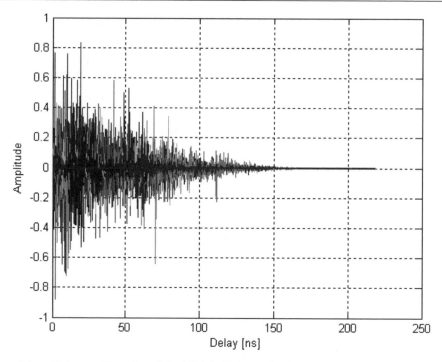

Figure 2.15   Delay profiles of modified Saleh–Valenzuela channel 3, 100 channel realizations

was 2–8 GHz which was covered with 1601 frequency points. Transmitted power measured at the input of transmitter antenna at 2 GHz was $P_{TX} = +10\,\mathrm{dBm} \pm 1\,\mathrm{dB}$ and the typical antenna gain was 0 dBi. Based on the maximum noticeable delay definition and the used parameters, the reflections exceeding the delay $\tau_m = 1600/$ (6 GHz) = 266.6 ns, which corresponds to 80 m, cannot be detected.

Figures 2.18 and 2.19 show the measured impulse responses as a snapshot given by a VNA which has a TD-measurement option available. The delay grid is 10 ns and the power scale grid is 10 dB. Because the complex samples can be recorded when the original VNA measurement system is used, the propagation delay, and therefore also the link distance, can be calculated from the results, as can be seen from the example figures.

The room sizes examined are 7430 mm × 4100 mm and 5940 mm × 6280 mm for the results presented in Figures 2.18 and 2.19, respectively, and the heights are typically around 3.5 m. In Figure 2.18, the 7.438-ns delay for the first detected path corresponds to a distance of 2.23 m. The room where the measurements were carried out is the one presented in Figure 2.9 (right). The 40-dB dynamic range gives the 53 ns and 23 ns delay spreads for the given examples. However, for the communication applications, the maximum 10-dB dynamic range limits the exploitable multipaths for Rake receiver as is the case also in IEEE802.15.3 model definition (Foerster, 2003; Foerster *et al.*, 2003).

The following results have been calculated for antenna heights $h_{TX} = 2.2\,\mathrm{m}$ and $h_{RX} = 1.1\,\mathrm{m}$, and vice versa (Hovinen *et al.*, 2002). The link separations in this data set were between 1.25 m and 8.10 m, which corresponds to initial propagation delays from 4.2 ns to 27.0 ns. The first measurement campaign covered situations where the

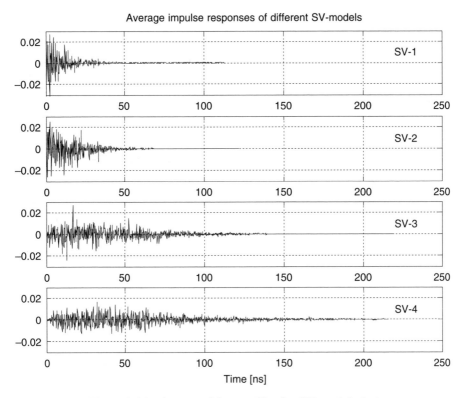

Figure 2.16   Average delay profiles for SV-models 1–4

transmitting and receiving antennas were located in the same room. Both the LOS and NLOS links were studied. The measurements where TX-antenna and RX-antenna were in the different rooms (through-wall measurement) were also carried out.

The amplitude response data were inverse Fourier transformed into delay domain. A Hanning window was applied before transformation to make it easier to locate the line-of-sight component of the signal. As phase information was available, the propagation delay was able to be observed in the measured data. The data was simultaneously transformed into delay domain using no windowing in order to avoid data manipulation. The system noise level was estimated from the absolute delay range 0–3 ns which covers the time of flight, and was found to be around −105 dBm, typically, limiting the dynamic range to 40 dB. The time-of-arrival (TOA, initial delay) of the LOS component was extracted for each of the radio links using the average delay profile of 500 impulse responses (small-scale statistics). Initial delay was removed from the results, and the statistical parameters were extracted from this data.

The maximum excess delay was limited to 70 ns, which corresponds to 420 samples in delay domain. The limit was found by removing the strongest reflections and plotting average delay profiles from various data sets. This corresponds to large-scale statistics represented in Figure 2.20, where the reflections are plotted in dark grey tint.

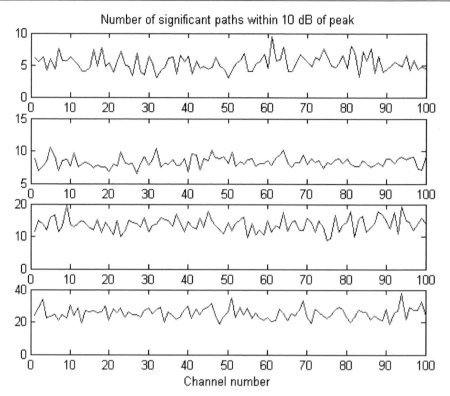

Figure 2.17 Number of significant paths within 10 dB dynamic range for different SV-models

A discrete model for the output of a time-variant fading multipath channel is given by

$$h(t) = \sum_n a_n(t)s(t - \tau_n(t)), \tag{2.12}$$

where $s(t)$ is the transmitted signal, $a_n(t)$ is the amplitude gain of the nth multipath channel and $\tau_n(t)$ is the corresponding excess delay. In an indoor environment with no moving scatterers and with fixed antenna positions, $a_n(t)$ and $\tau_n(t)$ are assumed to be constant during the observation time. Measurements recorded in various antenna positions give an estimate of the average static indoor channel. Movement of scatterers or altering the length of the radio link will introduce Doppler shift, which in turn reduces the coherence time t_{coh} of the channel. The channel can be measured if the measurement time is shorter than the coherence time. This assumption is valid in static indoor measurements.

A multipath channel is typically modelled as a linear tapped delay line (finite impulse response, FIR, filter), with complex tap coefficients (Bello, 1963). In computer simulation, the time variance of the channel filter is realized in various ways, the simplest and most straightforward being a FIR filter, whose coefficients are updated from previously stored complex-valued channel data. The drawback of this approach is the limited randomness, since the data eventually has to be reused. Statistical channel description

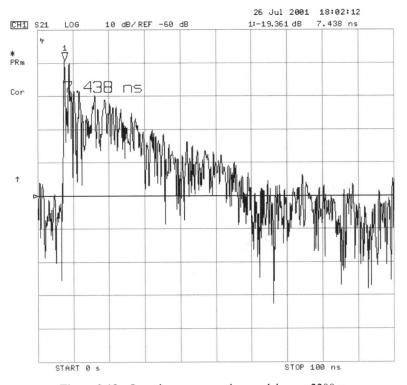

Figure 2.18   Impulse response, h_{TX} and $h_{\mathrm{RX}} = 2200\,\mathrm{m}$

gives more freedom for simulation. In the tapped delay line model, the time variation is realized by mixing the multiplicative white noise through a bandpass filter directly to the tap coefficients.

Nevertheless, a realistic assumption for a static indoor channel is a Ricean fading model. Ricean fading signals have amplitude $a_n(t)$ that is distributed according to (Proakis, 1995)

$$p_R(a) = \frac{a}{\sigma^2} \exp\left(-\frac{a^2 + s^2}{2\sigma^2}\right) I_0\left(\frac{as}{\sigma^2}\right), a \geq 0 \qquad (2.13)$$

where σ is the standard deviation and I_0 is the zeroth order modified Bessel function of the first kind. The non-centrality parameter s is defined by

$$s^2 = \|\bar{a}(t)\|^2 \qquad (2.14)$$

where $\bar{a}$ is mean complex amplitude.

k-factor of a Ricean fading signal is defined as

$$k = \frac{s^2}{2\sigma^2} = \frac{s^2}{\eta^2} \qquad (2.15)$$

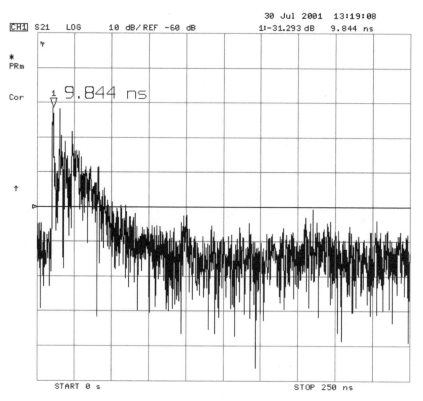

Figure 2.19   Impulse response, $h_{TX} = 2200\,\text{mm}$ and $h_{RX} = 600\,\text{mm}$

which in logarithmic scale is

$$k = s_{dB}^2 - \eta_{dB}^2. \tag{2.16}$$

Rayleigh fading channel is a special case of Ricean channel with $k = 0$. It has been shown (Talvitie, 1997) that the Ricean fading channels become effectively Rayleigh fading when k becomes smaller than 5 dB.

A large-scale (long-term) model is constructed from data that has been collected in various locations. Because antenna positions and room sizes vary, we need to separate pure reflections from the random part of the model. In other words, the channel model contains a deterministic environment-dependent ray-tracing part (deterministic model, DM) and a statistical environment-independent Ricean fading part (statistical model, SM) as presented in Figure 2.21.

The DM contains static reflections, delay and power estimates of which can be modeled either with ray tracing tools or calculated from a simplified reflection model. The power of the LOS signal is normalized to unity, and it is contained in the DM part.

Parameters for SM are extracted from the measurement data after normalizing to average power level and removing the most significant reflections. Figure 2.22 (a–b)

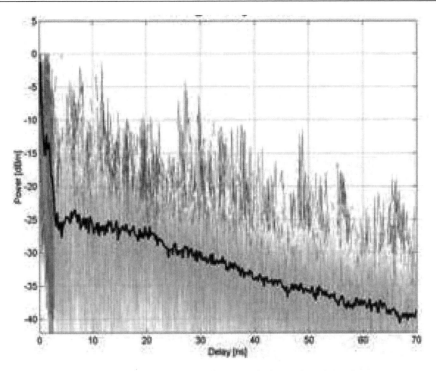

Figure 2.20   Average delay profile of 32 500 independent impulse responses

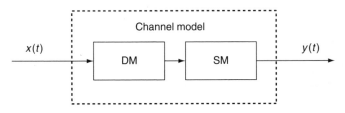

Figure 2.21   Channel model blocks

shows the statistical parameters of SM as functions of antenna separation at delays 0 ns
and 5 ns, respectively. In other words, k decreases exponentially in linear scale. Estimates for $k(\tau, d)$ can be tabulated directly, but here they have been expressed using
estimates of signal power $s^2(\tau, d)$ and noise power (τ, d). Figure 2.23 (a–b) shows the
values for regression parameters a_s, a_η and b_s, b_η as functions of excess delays. Again, we
can fit piecewise linear regression lines to this data and get the following expressions

$$a_s(\tau) = \alpha'_s \tau + \alpha_s$$

$$a_\eta(\tau) = \alpha'_\eta \tau + \alpha_\eta$$

$$(2.17)$$

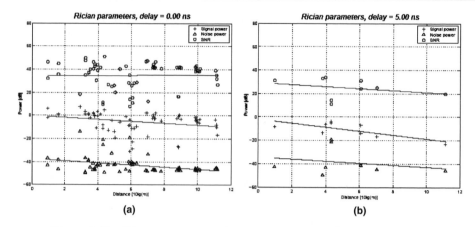

Figure 2.22   Ricean parameters for line-of-sight path and at delay 5 ns

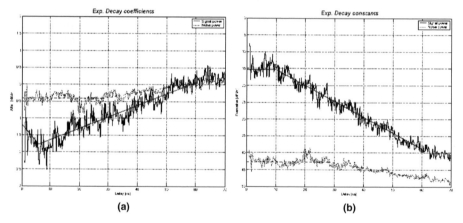

Figure 2.23   Coefficients (a) a_s and a_η and (b) b_s and b_η for exponential decay of Ricean parameters

and

$$b_s(\tau) = \beta'_s \tau + \beta_s,$$
$$b_\eta(\tau) = \beta'_\eta \tau + \beta_\eta \qquad (2.18)$$

where subscripts s and η are associated to $s^2(d, \tau)$ and to $\eta^2(d, \tau)$, respectively. Table 2.2 and Table 2.3 list the regression parameters for the signal power and the noise power.

The signal power and the noise power can be expressed in compact matrix forms

$$\mathbf{S} = \mathbf{a}_s^{\mathrm{T}} \mathbf{d} + \mathbf{b}_s^{\mathrm{T}},$$
$$\mathbf{N} = \mathbf{a}_\eta^{\mathrm{T}} \mathbf{d} + \mathbf{b}_\eta^{\mathrm{T}} \qquad (2.19)$$

Table 2.2 Parameters for exponential decay coefficient α

Delay range	Signal power		Delay range	Noise power	
τ	α'_s	α_s		α'_η	α_η
0–6 ns	−0.095	−1.20	0–28 ns	0.0	−0.42
6–60 ns	0.035	−1.98	28–60 ns	0.015	−0.84

Table 2.3 Parameters for exponential decay constant β

Delay range	Signal power		Delay range	Noise power	
τ	β'_s	β_s		β'_η	β_η
0–10 ns	0.00	−15.2	0–28 ns	0.00	−42.7
10–60 ns	−0.47	−10.6	28–60 ns	−0.15	−38.5

where T denotes transpose operation, and dimensions of **a** and **b** depends on delay, and dimension of **d** depends on antenna separation. Signal-to-noise ratios can in dB be expressed as

$$\mathbf{K} = \mathbf{S} - \mathbf{N}. \tag{2.20}$$

Matrices **S**, **N** and **K** contain the signal power (delay profile), noise power and the SNR values for Ricean fading channel, respectively. Each of the columns describes the fading statistics as a function of delay at the given antenna separation d. When K is below 5 dB, a Rayleigh channel with corresponding noise power applies (Talvitie, 1997).

The channel model presented can be applied to computer simulations for performance studies of UWB systems.

The impulse responses based on the indoor measurements carried out at the University of Oulu also support the clustered ray arrival phenomenon presented in Chapter 3.1. UWB radio channel characteristics have also been studied lately (for example, Kunisch and Pamp, 2002; Cassioli et al., 2002; Gramer et al., 2002).

2.4 Path Loss Models

The measurement data collected in a series of multipath propagation studies can also be used to model UWB path loss. The impact of the link distance on the received signal energy can be determined by propagation loss calculations which define the fraction of the transmitted power that can be received at the distance d. General propagation physics approaches are valid in the case of UWB transmission, which means that the longer the link distance is, the lower is the received signal energy. This connection can easily be derived by the well known path loss.

$$PL(d) \approx \left(\frac{d}{d_0} \right)^{-n} \tag{2.21}$$

where d is the link separation, d_0 is the reference distance that is usually 1 m, and n is environment dependent constant. In free-space, $n = 2$.

At the University of Oulu premises, path-loss focused measurements have been carried out using the VNA measurement system from Figure 2.8. The received signal power as a function of distance was studied in two different sized corridors and a lecture hall.

Figure 2.24 and Figure 2.25 present the layouts and dimensions for the corridors where the soundings were carried out. The frequency range was 2–8 GHz, the antenna heights were 110 mm. The heights of corridors 1 and 2 were 3890 and 2280 mm, respectively. The height of the lecture room was 3540 mm.

The soundings were made at the lecture hall at distances between the 152 cm at minimum and 1052 cm at maximum, see Figure 2.26.

The received signal powers were studied as a function of the distance, which indicates the attenuation of the signal. The exponential correspondence can be noticed in an absolute

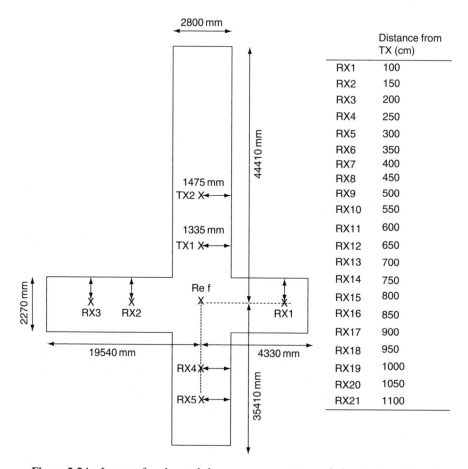

Figure 2.24 Layout for the path loss measurements carried out at corridor 1

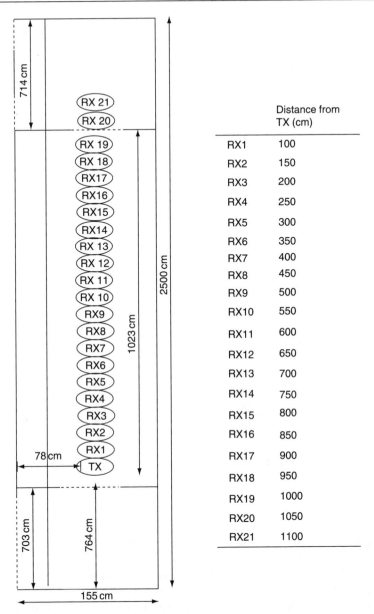

	Distance from TX (cm)
RX1	100
RX2	150
RX3	200
RX4	250
RX5	300
RX6	350
RX7	400
RX8	450
RX9	500
RX10	550
RX11	600
RX12	650
RX13	700
RX14	750
RX15	800
RX16	850
RX17	900
RX18	950
RX19	1000
RX20	1050
RX21	1100

Figure 2.25 Layout for the path loss measurements carried out at corridor 2

distance scale. The attenuation corresponds to the straight line in a logarithmic scale. A linear regression line is then fitted to the measured power data points using the equation

$$P(d) = n10\log_{10}(d) + a \qquad (2.22)$$

where n corresponds to the path loss factor, d is distance and a is a power scaling constant.

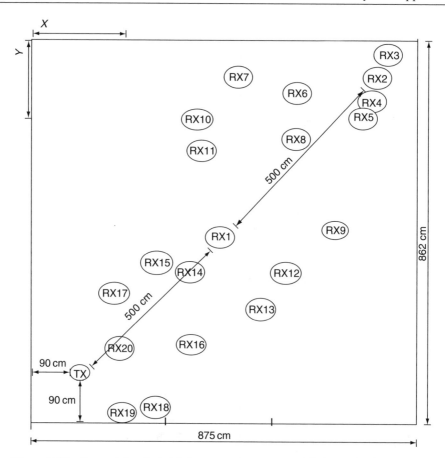

Figure 2.26 Layout for the path loss measurements carried out at a lecture hall

There are two possible ways of fitting the regression line to the data points. The more accurate method is to use piecewise fitting where the path loss factor is different for the shorter distances than the longer distances. The other way is to use only one regression line. Figure 2.27 presents the example of the curve fitting. In this example, the piecewise fitting matches the data points almost perfectly, but the single line gives the general slope for the line. The locations of the measured data points depend on the environment, the materials used and the link parameters.

Table 2.4 collates the final results based on the various path loss measurements. In addition to the LOS measurements, a link through the wall was measured using different antenna heights. The measurements showed that the UWB signal attenuation is smaller than in freespace when the propagation media includes the line-of-sight component.

Figure 2.28 presents the measurement results and the fitted linear regression lines for the NLOS data. The soundings were done between two rooms that were

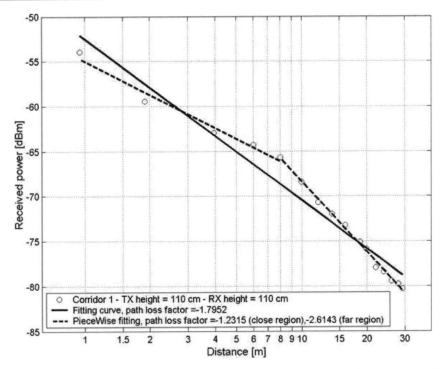

Figure 2.27   Linear regression lines for the path loss measurements carried out at corridor 1

separated by the corridor. The link distances were from 3 m to 13 m. The transmitting antenna height was set to 220 cm. The measurements were carried out using three different receiver antenna heights: 60, 110 and 220 cm. When the signal was obstructed by walls, the signal was also significantly attenuated leading to reduced received power.

2.5   Conclusions

This chapter has explored some common channel measurement techniques as well as presenting commonly used channel models.

Table 2.4   Path loss factors for different environments

Location	TX height (cm)	RX height (cm)	k
Lecture hall	110	110	-1.0454
Corridor 1	110	110	-1.7952
Corridor 2	110	110	-1.4386
Through wall (NLOS)	220	60	-3.8514
Through wall (NLOS)	220	110	-3.3009
Through wall (NLOS)	220	220	-3.1797

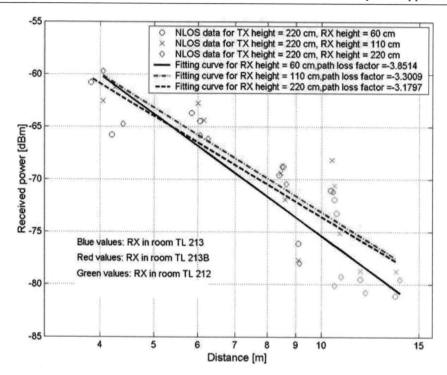

Figure 2.28 Path loss in NLOS measurements

Due to the large frequency band that the UWB signal occupies, the frequency diversity of the signal is extremely high. The UWB signal is therefore not very sensitive to deep notches in any part of the signal spectrum and the fading mechanism differs greatly in the opposite edges of the signal band.

Due to the large bandwidth of the UWB signal, the propagation phenomena are different in the lower and upper bands of the signal spectrum. The size of the Fresnel zone, which is a function of wavelength, will be markedly different at the lower and higher frequency ends. Surface roughness and reflection coefficients may also be significantly different as the wavelength varies significantly over the available signal bandwidth.

In general, the UWB channel model is very similar to well-known tap-delay line models for wideband systems. One characteristic that is different is the relatively large number of multipath components.

3

Modulation Schemes

Matti Hämäläinen, Raffaello Tesi, Ian Oppermann

3.1 Introduction

Many different pulse generation techniques may be used to satisfy the requirements
of an UWB signal. As discussed in the previous chapters, the FCC requires that
the fractional bandwidth is greater than 20 %, or that the bandwidth of the transmitted signal is more than 500 MHz, whichever is less. The FCC also stipulates peak
power requirements (Federal Communications Commission, 2002a) Many possible
solutions may be developed within these restrictions to occupy the available bandwidth.

UWB systems have historically been based on impulse radio concepts. Impulse radio
refers to the generation of a series of very short duration pulses, of the order of
hundreds of picoseconds. Each pulse has a very wide spectrum, which must adhere to
the spectral mask requirements. Any given pulse will have very low energy because of
the very low power levels permitted for typical UWB transmission. Therefore, many
pulses will typically be combined to carry the information for one bit. Continuous pulse
transmission introduces a complication in that, without further signal processing at the
transmitter, strong spectral lines will be introduced into the spectrum of the transmitted
signal. Several techniques are available for minimizing these spectral lines, the most
common of which are described later in this chapter.

Impulse radio has the significant advantage of being essentially a baseband technique.
The most common impulse radio based UWB concepts are based on pulse position
modulation with time hopping (TH-PPM). Time hopping, direct sequence techniques
and multi-carrier schemes are described in this chapter. However, the focus will be on
impulse radio modulation schemes.

UWB Theory and Applications   Edited by I. Oppermann, M. Hämäläinen and J. Iinatti
© 2004 John Wiley & Sons, Ltd   ISBN: 0-470-86917-8

3.2 Impulse Radio Schemes

3.2.1 Impulse radio UWB

Time-modulated ultra wideband (TM-UWB) communication is based on discontinuous emission of very short Gaussian pulses or other types of pulse waveforms (monocycles), as in Figure 3.1(a). Each pulse has the ultra wide spectral requirement in frequency domain like that in Figure 3.1(b). This type of transmission does not require the use of additional carrier modulation as the pulse will propagate well in the radio channel. The technique is therefore a baseband signal approach. This radio concept is referred as impulse radio (IR).

One transmitted symbol is spread over N monocycles to achieve processing gain that may be used to combat noise and interference. This is similar to the approach used for spread spectrum systems. The processing gain in dB derived from this procedure can be defined as

$$PG_1 = 10 \log_{10}(N). \tag{3.1}$$

The monocycle waveform can be any function which satisfies the spectral mask regulatory requirements. Common pulse shapes include Gaussian, Laplacian, Rayleigh or Hermitian pulses. Data modulation is typically based on pulse position modulation (PPM). Conroy *et al.* (1999) present pulse amplitude modulation (PAM) in UWB transmission. The UWB receiver is a homodyne cross-correlator that is based on the architecture that utilizes a direct RF-to-baseband conversion. Intermediate frequency conversion is not needed, which makes the implementation simpler than in conventional (super-)heterodyne systems.

Unlike spread spectrum systems, the pulse (chip) does not necessarily occupy the entire chip period. This means that the duty cycle can be extremely low. The receiver is only required to 'listen' to the channel for a small fraction of the period between pulses. The impact of any continuous source of interference is therefore reduced so that it is only relevant when the receiver is attempting to detect a pulse. This leads to processing gain in the sense of a spread spectrum system's ability to reduce the impact of interference. Processing gain due to the low duty cycle is given by

$$PG_2 = 10 \log_{10}\left(\frac{T_f}{T_p}\right), \tag{3.2}$$

where T_f is time hopping frame and T_p is impulse width.

Total processing gain PG is a sum of the two processing gains (Time Domain Corporation, 1998):

$$PG = PG_1 + PG_2. \tag{3.3}$$

As the transmitted signal is not continuous, UWB communication is resistant to severe multipath propagation. If the time between pulses is greater than the channel delay spread, there is no ISI between pulses and so no ISI between bits. In discontinuous

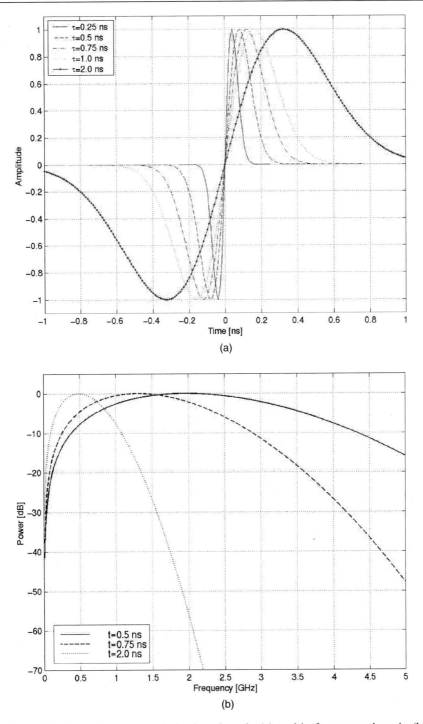

Figure 3.1 Gaussian monocycle in time domain (a) and in frequency domain (b)

transmission, consecutive pulses are sent within a time frame (T_f) which is defined by a Pseudo random (PR) time-hopping code. Due to the short pulse width and a relative long pulse repetition time (compared with pulse width), the transmitted pulse is attenuated before the next pulse is sent. This reduces interpulse interference.

In the time domain, the transmitted Gaussian monocycle can be defined as the first derivative of the Gaussian function. Figure 3.1 shows the time and frequency domains of sample monocycles of different duration.

The Gaussian monocycle in time domain $v(t)$ can be mathematically defined using the formula (Time Domain Corporation, 1998)

$$v(t) = 6A\sqrt{\frac{e\pi}{3}}\frac{t}{T_p}e^{-6\pi(t/T_p)^2},\tag{3.4}$$

where A is pulse amplitude, T_p is pulse width and t is time.

The corresponding function in the frequency domain is the Fourier transform of $v(t)$, which can be defined as

$$F\{v(t)\} = V(f) = -j\frac{Af\,T_p^2}{3}\sqrt{\frac{e\pi}{2}}\,e^{-\frac{\pi}{6}f^2T_p^2},\tag{3.5}$$

where f is frequency.

The nominal centre frequency and the bandwidth of the monocycle depends on the monocycle's width. The $-3\,\mathrm{dB}$ bandwidth is approximately 116 % of the monocycles nominal centre frequency $f_0 = 1/T_p$ (Time Domain Corporation, 1998). Considering an example monocycle duration of 0.75 ns (as shown in Figure 3.1(a)), the nominal centre frequency is 1.33 GHz and the half power bandwidth is 1.55 GHz. The spectrum of Gaussian monocycle is asymmetrical, as can be seen in Figure 3.1(b).

The ideal Gaussian monocycle has a single zero crossing. If additional derivatives of the Gaussian pulse are taken, the relative bandwidth decreases, and the centre frequency increases for a fixed time decay constant T_p. Additional zero crossings of the pulse will reduce the space and time resolution of the system due to the decreased bandwidth.

Figure 3.2 presents a block diagram of a time-hopping UWB impulse radio concept that utilizes pulse position modulation (early concept by Time Domain Corporation, USA).

The pulse waveforms and their spectra are collected in Figures 3.3 and 3.4. In Figure 3.3, the solid line represents the generated pulse, and the dashed line represents the pulse waveform in the channel. The UWB antenna acts as a high pass filter, and may be thought of as a differentiation block in the time domain (Ramirez-Mireles and Scholtz, 1998a). The transmitted pulse is therefore the first derivative of the generated pulse waveform.

In Figures 3.3 and 3.4, a Gaussian doublet is also presented as a potential waveform candidate. However, this waveform is more suitable for geolocation and positioning applications than for communication applications because of the longer total bi-pulse

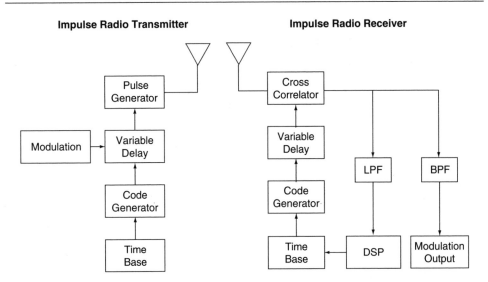

Figure 3.2 Block diagram of the TH-PPM UWB impulse radio concept presented by Time Domain Corporation

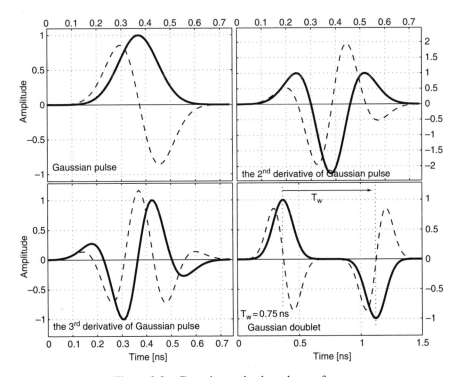

Figure 3.3 Gaussian-pulse-based waveforms

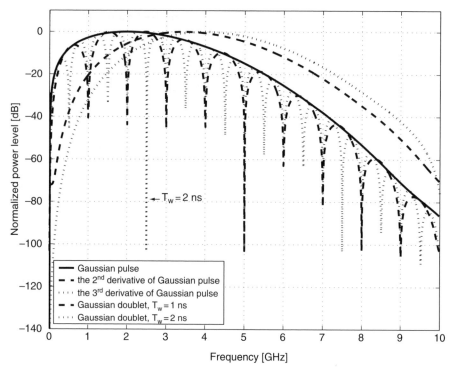

Figure 3.4 Spectra of the pulse waveforms used (pulse width 0.5 ns)

width. The doublet consists of two amplitude reversed Gaussian pulses with a time delay T_w between the pulses. This limits the use of the doublet in high speed data communication applications. In specific cases, T_w can be used to generate nulls to the spectrum in order to avoid intentional interference. This can be seen from Figure 3.4.

3.2.2 Fast Stepped Frequency Chirps

Both the bandwidth and spectral mask limitations must be satisfied by the UWB signal. There are many means of generating a very wideband signal. One technique is based on fast frequency chirps, which are commonly used in impulse radar applications. It is possible to generate a wideband transmission by sweeping the transmitter's oscillator in the frequency domain. A bandwidth of several hundred MHz can be achieved with ~10 ns sweep time (Stickley *et al.*, 1997). Wider bandwidths can be achieved using this technique. For example, ground penetrating radar (GPR) systems with 50–1200 MHz bandwidth have been documented (Carin and Felsen, 1995).

GPRs based on UWB technology are suitable for object detection in, for example, landmine sweeping and avalanche rescue operations because of the good signal penetration ability and fine space resolution.

3.3 Multi-Carrier Schemes

3.3.1 Multi-carrier Spread Spectrum Schemes

Another approach is to extend the techniques utilized for direct sequence spread spectrum or code division multiple access (CDMA) schemes that are used for third-generation mobile systems. Wideband CDMA systems with optional multi-carrier (MC) techniques can be used to fill the available spectral mask.

There are three main techniques for generating a SS-MC transmission: multi-carrier CDMA, multi-carrier DS-CDMA, and multitone (MT) CDMA (Prasad and Hara, 1996). Each of these techniques relies on, and benefits from, the properties of conventional spread spectrum signals. However, multi-carrier systems are reasonably complex to implement. In particular, multi-carrier systems require several mixers or digital fast Fourier (FFT) transform techniques to place the different signal components in the required bands.

Figure 3.5 shows a block diagram for a MC-CDMA system. The original data stream is spread over the different sub-carriers f_i with each chip of pseudo random (PR) code C_i. The spectrum spreading is done in the frequency domain (Prasad and Hara, 1996). The signal is de-spread at the receiver using corresponding chips g_i of the spreading code C_i. In UWB applications, the individual modulated carrier needs to fulfil the 500 MHz bandwidth requirements.

Figure 3.6 shows a block diagram for a multi-carrier DS-CDMA system. This method involves spreading the original data in the time domain after serial-to-parallel conversion of the data stream (Prasad and Hara, 1996). The system needs to obtain the 500 MHz minimum bandwidth requirement to be treated as UWBs.

Figure 3.7 shows the block diagram of a multi-tone CDMA system. The bandwidth of the MT-CDMA system is smaller than in the previous multi-carrier systems because of the small sub-carrier spacing. This kind of multi-carrier approach is the closest to the original UWB idea. However, it causes the highest self-interference due to the overlapping spectra.

Multi-carrier technology is currently used in high data rate applications, for example in WLAN systems such as Hiperlan2, Digital Audio or Video Broadcasting (DAB and DVB, respectively) and in asymmetric digital subscriber line (ADSL).

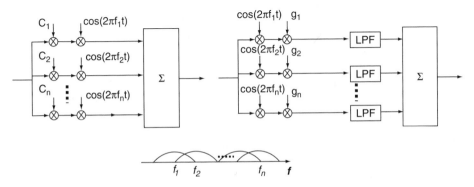

Figure 3.5 Block diagram and spectrum for multi-carrier CDMA system

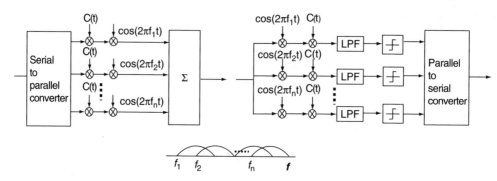

Figure 3.6   Block diagram and spectrum for multi-carrier DS-CDMA system

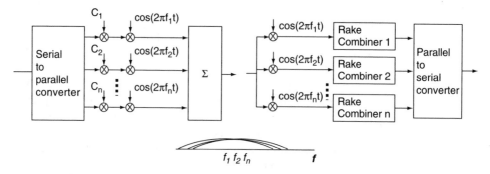

Figure 3.7   Block diagram and spectrum for multi-tone- CDMA system

The advantage of multi-carrier technologies over single carrier systems is that the data rate in each sub-carrier is lower than for single carrier systems. This eases synchronization of spreading sequence at the receiver, and helps to avoid ISI. The disadvantage is the increased complexity of the receiver requiring either multiple mixing stages or fast Fourier transform processing.

MC-CDMA schemes spread the signal in the frequency domain. The signal consists of overlapping, relatively narrow, carriers which fill the available UWB signal spectrum. MC-DS-CDMA and MT-CDMA schemes apply spreading in the time domain. MT-CDMA has a similar bandwidth to a basic DS-CDMA scheme (Prasad and Hara, 1996) with relatively small separation of the tones used (f_1 to f_n in Figure 3.7). Consequently, the spreading factor employed for MT-CDMA schemes are much higher than for the MC schemes. The increased spreading factor increases the processing speed required at the receiver.

The conventional DS-SS technique without multi-carrier properties can also be characterized as an UWB technique if the chip rate is high enough. This calls for extremely fast digital signal processing which may be impractical. Issues such as synchronization will also be a significant challenge.

3.3.2 *Multiband UWB*

A current proposal within the IEEE 802.15.3 working group for UWB signals (IEEE, 2004) utilizes overlapping groups of UWB signals which each have a bandwidth of approximately 500 MHz. This so called multiband UWB ensures adherence to the FCC minimum bandwidth requirements and allows efficient utilization of the available spectrum.

Figure 3.8 shows the spectrum plan for the first group of UWB signals. The sub-bands are spaced 470 MHz apart, and any number of 500 MHz signals may be utilized. This allows for flexible coexistence with existing communications systems (such as WLAN systems). Each sub-band is generated by an OFDM symbol with 10 dB bandwidth of ∼520 MHz. Figure 3.9 shows the second group of UWB signals, which overlap the first group of UWB signals by 235 MHz. This potentially enhances the system's flexibility with respect to coexistence, interference mitigation and multiple access.

Each group of UWB signals is divided into lower sets (sub-bands 1–8) and upper sets (sub-bands 9–15). Only seven sub-bands are used in the lower set, which means one sub-band can be avoided for coexistence. The upper set is used in parallel with the lower set to increase the bit-rate.

3.4 Data Modulation

A number of modulation schemes may be used with UWB systems. The potential modulation schemes include both orthogonal and antipodal schemes.

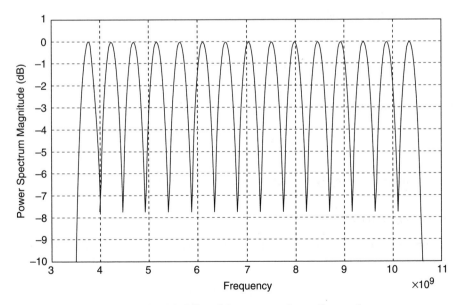

Figure 3.8 Multiband frequency plan—Group A

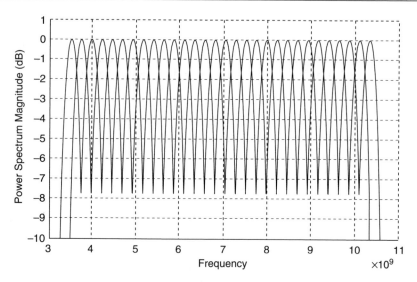

Figure 3.9 Multiband frequency plan—Group A and Group B

3.4.1 Pulse Amplitude Modulation

The classic binary pulse amplitude modulation (PAM) can be presented using e.g. two antipodal Gaussian pulses as shown in Figure 3.10. The transmitted binary baseband pulse amplitude modulated information signal (*t*) can be presented as

$$x(t) = d_j \cdot w_{tr}(t), \tag{3.6}$$

where $w_{tr}(t)$ represents the UBW pulse waveform, j represents the bit transmitted ('0' or '1') and

$$d_j = \begin{cases} -1, & j = 0 \\ 1, & j = 1 \end{cases}. \tag{3.7}$$

In Figure 3.10, the first derivative of the Gaussian pulse is shown, analytically defined as

$$w_{G_1}(t) = -\frac{t}{\sqrt{2\pi}\sigma^3}\, e^{\left(-\frac{t^2}{2\sigma^2}\right)}, \tag{3.8}$$

where deviation σ is directly related with the pulse length T_p by $\sigma = T_p/2\pi$.

3.4.2 On–Off Keying

The second modulation scheme is binary on–off keying (OOK). Using the following definitions

$$d_j = \begin{cases} 0, & j = 0, \\ 1, & j = 1 \end{cases} \tag{3.9}$$

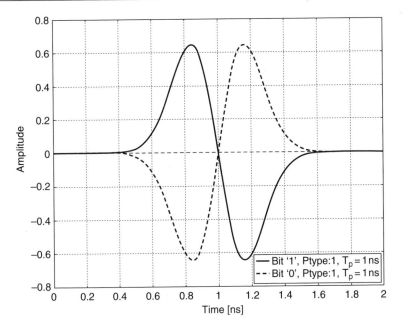

Figure 3.10 BPAM pulse shapes for '1' and '0' bits

the waveform used for this modulation can be defined in equation 3.8 (Figure 3.11). The major difference between OOK and PAM is that nothing is transmitted in OOK when bit '0' is chosen.

3.4.3 Pulse Position Modulation

With pulse position modulation (PPM), the chosen bit to be transmitted influences the position of the UWB pulse. That means that while bit '0' is represented by a pulse originating at the time instant 0, bit '1' is shifted in time by the amount of δ from 0. Analytically, the signal can be represented as

$$x(t) = w_{tr}\left(t - \delta d_j\right), \tag{3.10}$$

where d_j assumes the following values, depending on the bit chosen to be transmitted,

$$d_j = \begin{cases} 0, & j = 0 \\ 1, & j = 1, \end{cases} \tag{3.11}$$

and the other variables have been defined previously.

The value of δ could be chosen according to the autocorrelation characteristics of the pulse. The autocorrelation function can be analytically defined as (Proakis and Salehi, 1994)

$$\rho(t) = \int_{-\infty}^{+\infty} w_{tr}(\tau)w_{tr}(t - \tau)d\tau. \tag{3.12}$$

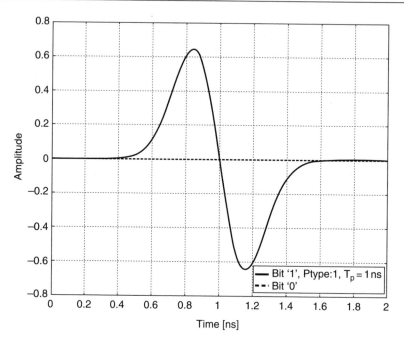

Figure 3.11 OOK pulses used for '1' and '0' bits

For instance, if we want to implement a standard PPM with orthogonal signals, the optimum value for δ (which we call δ_{opt}) will be the one which satisfies

$$\rho(\delta_{opt}) = \int_{-\infty}^{+\infty} w_{tr}(\tau)w_{tr}(\delta_{opt} - \tau)d\tau = 0. \tag{3.13}$$

Figure 3.12 shows a particular case of PPM transmission where data bit '1' is sent delayed by a fractional time interval $\delta < 1$, and data bit '0' is sent at the nominal time.

The optimal modulation changes by using different pulse waveforms. The theoretical performance in an additive white Gaussian noise channel can be achieved with non-overlapping, orthogonal pulses, specifically, the modulation index $\delta \geq T_p$. However, optimal bit error rate (BER) performance and higher data rates will be achieved if the modulation index $\delta < T_p$ is as shown in Figure 3.13. The optimal modulation index is independent of pulse width because of the definition of relative fraction of pulse width.

As the order of the derivative increases, the minimum bit error rate is reached for a lower value of δ, and better BER performance is achieved. The justification for this behaviour is related to the cross-correlation of the pulses related to the data-bits '0' and '1'. Figure 3.14 shows the autocorrelation values for different kinds of pulses. As previously mentioned, the pulse width does not affect the results related to the different values of δ.

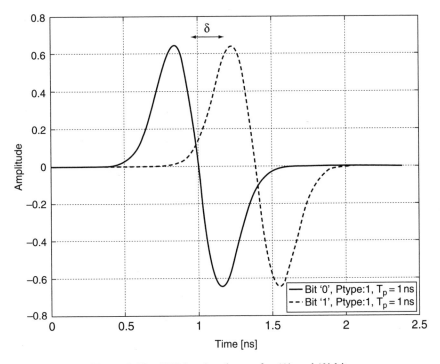

Figure 3.12 PPM pulse shapes for '1' and '0' bits

Two peculiarities should be noted for PPM modulation:

- The autocorrelation functions of the Gaussian waveforms have both positive and negative values. This explains why it is possible to achieve a better BER performance than the BER performance for time-orthogonal pulses with δ values less than 1 ($\delta \geq 1$ implies time-orthogonal signals, as seen in Figure 3.14).
- The autocorrelation minima occur at δ values, which correspond to the best BER performances.

The behaviour of the cross-correlation provides a means of selecting the optimal value of δ for the AWGN channel case. The value of δ can be fixed *a priori* once the UWB pulse waveform has been chosen. The best value to use for δ can be selected once the cross-correlation of the selected pulse waveform is calculated. The optimal value of δ for each pulse waveform is presented in Table 3.1.

3.4.4 Pulse Shape Modulation

Pulse shape modulation (PSM) uses different, orthogonal waveforms to represent bit '0' and '1'. The transmitted pulse can be represented as

$$x(t) = \left(1 - d_j\right) w_{tr}^{(0)}(t) + d_j w_{tr}^{(1)}(t), \tag{3.14}$$

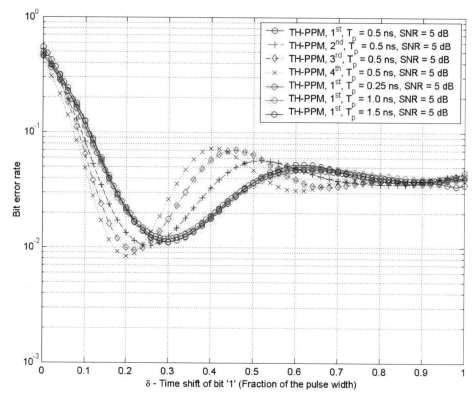

Figure 3.13   Bit error rate for different pulse waveforms as a function of δ

Table 3.1   Optimal time shift values δ for PPM modulation in an AWGN channel

Waveform	Optimal δ
Second Derivative	$0.292683 \cdot T_p$
Third Derivative	$0.243902 \cdot T_p$
Fourth Derivative	$0.219512 \cdot T_p$
Fifth Derivative	$0.195122 \cdot T_p$

where d_j is defined as (2.11) and $w_{tr}^{(0)}$ and $w_{tr}^{(1)}$ represent two different waveforms.

Figure 3.15 shows an example pair of PSM pulses. This case uses the first and the second derivatives of the Gaussian pulse. These two waveforms are orthogonal according to the definition given by the cross correlation of the two waveforms

$$\rho_c(t)|_{t=0} = \int_{-\infty}^{+\infty} w_{tr}^{(0)}(\tau) w_{tr}^{(1)}(t-\tau) d\tau = 0. \tag{3.15}$$

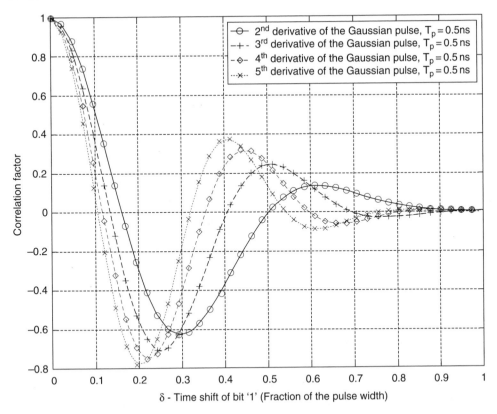

Figure 3.14   Autocorrelation of different pulse waveforms

3.4.5   *Theoretical Bounds*

The theoretical bounds in AWGN can be calculated for the modulation schemes presented. It is clear that BPAM is an antipodal modulation. Let $x_0(t)$ and $x_1(t)$ the signal representing bit 0 and 1 respectively. The BPAM signals are related by

$$\int_0^{T_p} x_0(t)x_1(t)dt = -1, \tag{3.16}$$

where T_p is the time duration of the signal. It can be shown that OOK and PPM are orthogonal using that same approach, that is,

$$\int_0^{T_p} x_0(t)x_1(t)dt = 0. \tag{3.17}$$

The orthogonality of the PSM signals depends on the pulse waveforms chosen to describe bit 0 and 1. For example, in the case of the first and second derivatives of the Gaussian pulse, it can be shown that the signals are orthogonal (da Silva and de Campos, 2002).

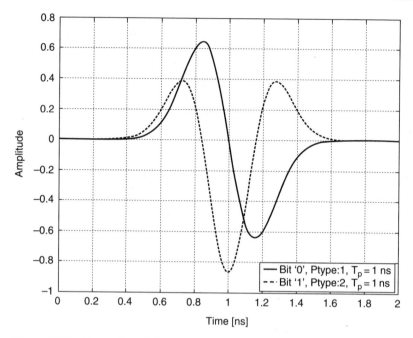

Figure 3.15 Examples of the pulse waveforms used for PSM modulation

The probability of error of an antipodal signal and of an orthogonal signal in an AWGN channel as a function of the signal-to-noise ratio has values (Proakis, 1995)

$$P_b^{(ant)} = Q\left(\sqrt{2 \cdot \text{SNR}}\right), \tag{3.18}$$

$$P_b^{(ort)} = Q\left(\sqrt{\text{SNR}}\right), \tag{3.19}$$

as depicted in Figure 3.16.

3.5 Spectrum 'Spreading'

Continuous pulse generation leads to strong spectral lines in the transmitted signal at multiples of the pulse repetition frequency. Data modulation typically occurs in a number of conceptual stages. First, a pulse train is generated. Second, a randomizing technique is applied to break up the spectrum of the pulse train. Third, the data modulation is applied to carry the information. The two main approaches to randomizing the pulse train are time hopping (TH) and direct sequence (DS) techniques.

The spectrum of a pulse train with and without randomizing techniques is depicted in Figure 3.17. Figure 3.17(a) shows the spectrum of a simple pulse train. As can be seen, the spectrum contains strong spectral lines at multiples of the pulse repetition frequency. The envelope of the spectrum is that of a single pulse. The regularity of these energy

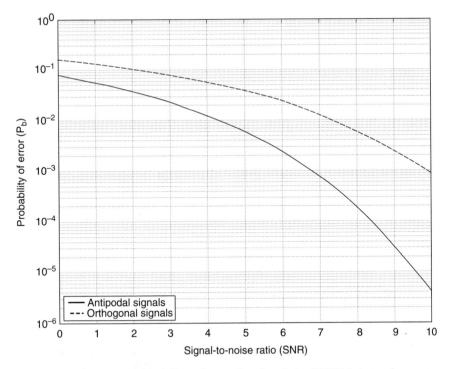

Figure 3.16 Probability of error for signals in AWGN channel

spikes may interfere with other communication systems at short range (Proakis, 1995). Randomizing the position in time of the generated pulses using data modulation and other randomizing techniques will affect the spectrum in such a way that the energy spikes are spread all over the spectrum, which is therefore smoothed. Figure 17(b) shows the spectrum of a pulse train that includes time hopping based randomization. A number of randomizing techniques may be found in the literature (Withington *et al.*, 1999). Randomizing is typically realized in UWB systems by a pseudo-random sequence.

3.5.1 TH-UWB

The data modulation is typically based on PPM using TH-UWB as the basis for a communication system. This approach allows matched filter techniques to be used in the receiver. Values of time shift (which is the modulation index for this form of modulation) have been reported as approximately one-quarter of a pulse width (Withington *et al.*, 1999). The optimum time shift depends on the cross-correlation properties of the pulses used. The concept of TH-PPM is presented in Figure 3.18 where only monocycle per data symbol is used (no processing gain is achieved). Figure 3.19 shows the construction of a single bit for TH-UWB systems.

TH-PPM monocycles spread the RF energy across the frequency band, reducing the large spikes in the pulse train spectrum. When a PR code is used to determine the transmission time within a large time frame, the spectra of the transmitted pulses

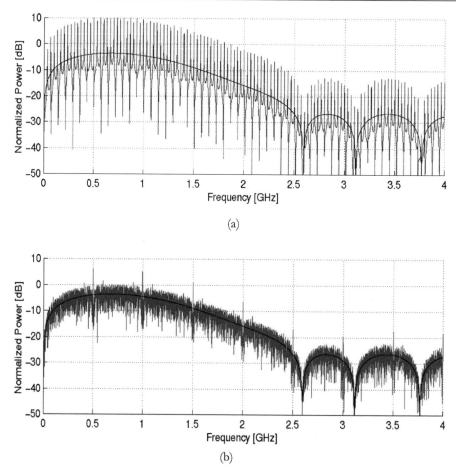

Figure 3.17 Spectrum of pulse train without (a) and with (b) randomizing techniques

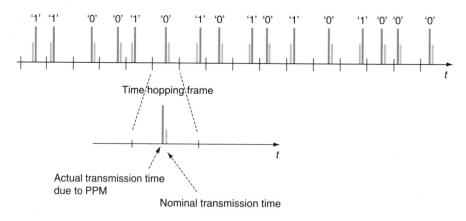

Figure 3.18 Time-hopping pulse position modulation technique

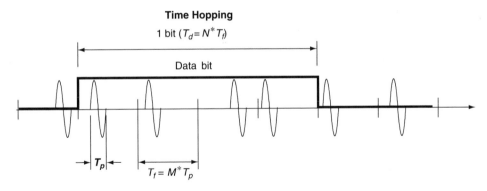

Figure 3.19 Time hopping system concept

become much more white-noise-like. The time hopping randomizes the signal in both time and frequency domains (Withington *et al.*, 1999). Pseudo random time hopping also minimises collisions between users in multiple access systems, where each user has a distinct pulse shift pattern (Win and Scholtz, 1997a).

However, a consequence of the PR time-modulation is that the receiver needs accurate knowledge of the PR code phase for each user. One can imagine impulse radio systems as time hopping spread spectrum systems. UWB waveforms are generated without any additional spreading. This simplifies the transceivers relative to conventional SS transceivers (Fontana *et al.*). The data rate of the transmission can be selected by modifying the number of pulses used to carry a single data bit (Kolenchery *et al.*, 1997). This in turn has an effect on the processing gain.

The pulse repetition time (or frame time) typically ranges from a hundred to a thousand times the pulse (monocycle) width. The symbol rate can be defined in terms of the number of monocycles used to modulate one data symbol in fixed frame time as

$$R_s = \frac{1}{T_s} = \frac{1}{N_s T_f} \, [\text{Hz}], \tag{3.20}$$

where R_s = symbol rate, T_s = symbol time, T_f = time hopping frame and N_s = number of monocycles/data bit.

If the data rate in (3.20) is reduced whilst the time hopping frame remains constant, the number of monocycles per data bit is increased. This leads to an increased processing gain.

3.5.2 Data Modulation with Time Hopping

In TH-mode, the pulse transmission instant is defined by the pseudo-random code. One data bit is spread over the multiple pulses to achieve a processing gain due to the pulse repetition (3.1). The processing gain is also increased by the low transmission duty cycle (3.2).

The TH spreading approach has been studied for PAM, PPM and PSM. However, OOK cannot take advantage of TH spreading because of the blank transmission in case of bit '0', and because it would create further problems for synchronization.

The information signal $s(t)$ for the mth user can be analytically described for PAM modulation as

$$s^{(m)}(t) = \sum_{k=-\infty}^{\infty} \sum_{j=0}^{N-1} w\left(t - kT_d - jT_f - (c_w)_j^{(m)} T_c\right) d_k^{(m)}, \qquad (3.21)$$

for PPM modulation as follows

$$s^{(m)}(t) = \sum_{k=-\infty}^{\infty} \sum_{j=0}^{N-1} w\left(t - kT_d - jT_f - (c_w)_j^{(m)} T_c - \delta d_k^{(m)}\right), \qquad (3.22)$$

and for PSM modulation as follows

$$s^{(m)}(t) = \sum_{k=-\infty}^{\infty} \sum_{j=0}^{N-1} w_{d_k^{(m)}}\left(t - kT_d - jT_f - (c_w)_j^{(m)} T_c\right). \qquad (3.23)$$

Figures 3.20, 3.21 and 3.22 show a single data bit for PAM, PPM and PSM respectively. Note that in the case of PAM and PPM the same TH sequence has been used to represent either data bit '0' and '1', while in the PSM case, two different sequences have

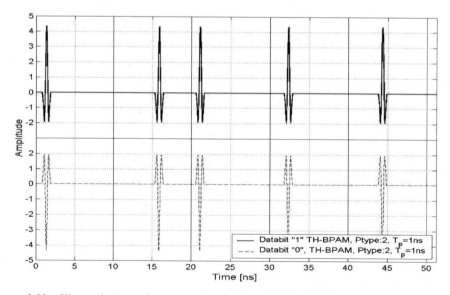

Figure 3.20   Time window of a transmitted data bit for **BPAM** modulation with TH spreading

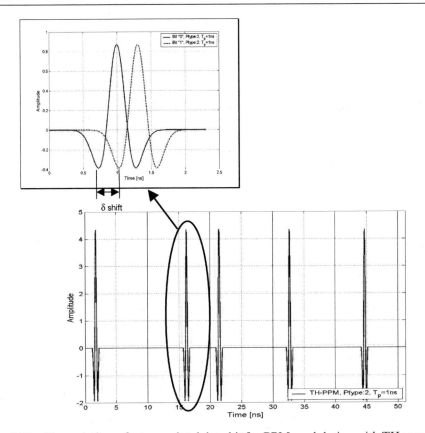

Figure 3.21   Time window of a transmitted data bit for PPM modulation with TH spreading

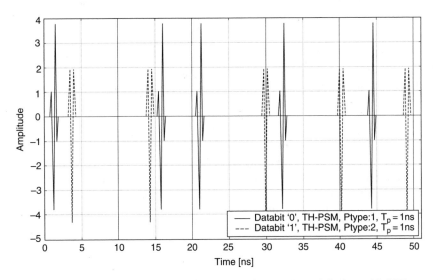

Figure 3.22   Time window of a transmitted data bit for PSM modulation with TH spreading

been considered. This results in a different allocation of the single pulses within the data bit frame. The pulse repetition frame is assumed to be 10 ns long in all of these figures.

3.5.3 Multiple Access with TH-UWB

In TH systems, users are separated using different PR codes of length N. In a frame, there are N possible transmission instants, so under ideal conditions a maximum of $M = N$ users can be allocated into the system without creating interference.

In time-hopping mode (TH-UWB), the modulated information signal $s(t)$ for the mth user can be written as: (3.21)

In TH-UWB $T_f \gg T_p$ producing a low duty cycle. In the following discussion, M is assumed to be the same as N. The input signal for the receiver in one user case is given by:

$$r(t) = \sum_{l=1}^{L} A_l s^{(1)}(t - \tau_l) + n(t) \tag{3.25}$$

where A_l is amplitude of radio channel path l, τ_l is delay of radio channel path l, L is the number of resolvable multipath components, $n(t)$ is additive white Gaussian noise.

The delay can be presented as a portion of T_c as $\tau_l = \xi T_c$. The effect of the antenna should be taken into account in the received signal waveform in $s(t)$.

Each pulse in a pulse train has a nominal transmission time, which is determined by the pulse repetition frequency (PRF). The actual transmission instant is varied from the nominal position by pulse position modulation. This pulse position modulation carries information, so that an early pulse represents a '0' and a delayed pulse represents a '1'. The determination of transmission for a given user is determined by the unique PR code.

The pulse repetition interval (which defines the length of each time hopping frame) is defined by the number of users multiplied by the length of a single time slot within the time hopping frame as given by

$$T_{PRF} = N_U T_C, \tag{3.26}$$

where T_C is the length of each time slot, and N_U is the number of users in the channel.

The maximum number of non-overlapping users is determined by the length of the PR code as seen in

$$N_U = 2^n - 1, \tag{3.27}$$

where n is the number of bits in the PN sequence generator.

The length of each of the time slots must be more than twice the length of a single pulse, since there must be enough time within the slot to transmit either a '0' or a '1'. This is assuming that the modulation index δ is no less than the length of a single pulse. This is assumed to be the minimum so as to avoid the overlapping of the pulses. In general, the length of a time slot is determined by

$$T_C > 2T_p + \delta, \tag{3.28}$$

where T_P is the length of the time delay (modulation index) used in PPM.

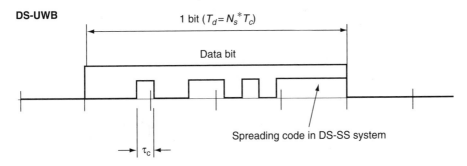

Figure 3.23   Direct Sequence system concept

For example, if the number of users is 31 and the pulse width is 800 ps, the length of the time slot within the pulse to be transmitted must be at least 1.6 ns. This results in a pulse repetition frequency of less than 21 MHz. By choosing the number of pulses per symbol to be 200, the processing gain is more than 41 dB.

3.5.4   Direct Sequence UWB

When utilizing DS techniques, a PR code is used to spread the data bit into multiple chips, much as in conventional DS spread spectrum systems. In the case of UWB systems, the pulse waveform takes the role of the chip in DS. The DS spreading approach has been studied for PAM, OOK and PSM modulation schemes. PPM modulation is intrinsically a time hopping technique since the bit value is given by the position of the pulse in a transmission slot. The use of the DS technique as a spreading approach would create a hybrid DS/TH configuration of the signal. Figure 3.23 shows the bit structure for a DS signal. The rectangular waveform indicates the individual chip elements.

The PAM and OOK information signal $s(t)$ for the mth user can be presented as

$$s^{(m)}(t) = \sum_{k=-\infty}^{\infty} \sum_{j=0}^{N-1} w(t - kT_d - jT_c)(c_p)_j^{(m)} d_k^{(m)}, \qquad (3.29)$$

where d_k is the kth data bit, $(c_p)_j$ is the jth *chip of the PR code*, $w(t)$ is the pulse waveform, N represents the number of pulses to be used per data bit, T_c is the chip length, The pseudo random code is bipolar assuming values $\{-1, +1\}$, The bit length is $T_d = NT_c = NT_p$.

3.5.4.1   Data Modulation with DS-UWB

Figure 3.24 shows a single data bit for binary PAM modulation when a data bit '1' or '0' is transmitted. The square wave represents the random code, which affects the polarity of individual pulses which make up the DS waveform. For clarity, only a small part of the data bit waveform is shown. Pulse type (Ptype) defines the pulse waveform used. Ptype2 is the second derivative of the Gaussian pulse.

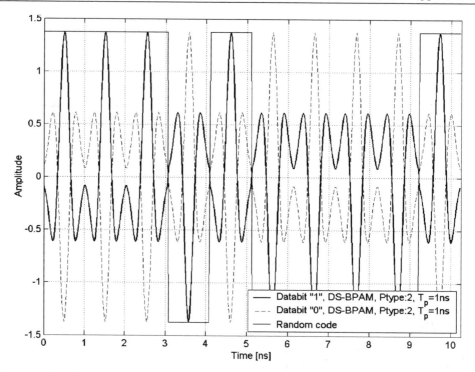

Figure 3.24 Time window of a transmitted data bit for BPAM modulation with DS spreading

Figure 3.25 shows a single data bit for OOK modulation when data bit '1' or '0' is transmitted. As mentioned above, in case of data bit "0", the OOK spreading signal is characterized by no transmission.

For PSM modulation, the information signal $s(t)$ for the mth user can be presented as

$$s^{(m)}(t) = \sum_{k=-\infty}^{\infty} \sum_{j=0}^{N-1} w_{d_k^{(m)}}(t - kT_d - jT_c)(c_p)_j^{(m)}, \qquad (3.30)$$

where the chosen bit d_k for the mth user, defined according to equation (3.14), determines the choice of the UWB pulse waveform to be transmitted.

Figure 3.26 shows a single data bit using for PSM modulation when data bit '1' or '0' is transmitted. Ptype1 and Ptype2 in the figure represent the first and the second derivatives of the Gaussian pulse.

3.5.5 Comparison of TH and DS BPAM UWB

This section examines the relative performance of BPAM for DS and TH techniques in an AWGN channel. The BER for up to 60 simultaneous users is examined. The

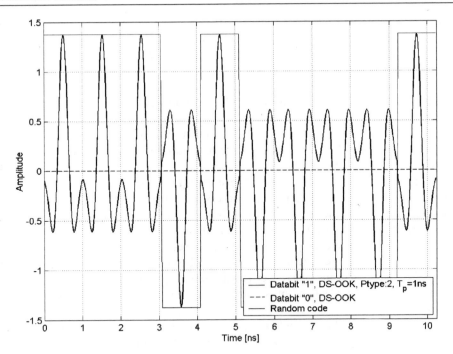

Figure 3.25 Time window of a transmitted data bit for OOK modulation with DS spreading

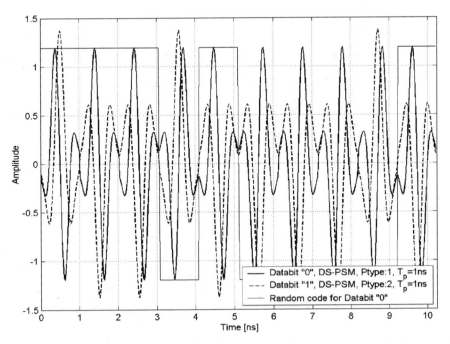

Figure 3.26 Time window of a transmitted data bit for PSM modulation with DS spreading

processing gain is 127 (approximately 21 dB). The DS system uses 127 pulses per bit, which is equal to the processing gain or the spreading sequence length, whilst the TH system uses 10 pulses per bit, which is the remainder of the processing gain coming from channel activity factor. Both synchronous and asynchronous transmissions are considered. All users have the same power. The pulse shape used is the fourth derivative of the Gaussian pulse. The pulse duration is 0.5 ns and the data rate is 16 Mbps.

Figure 3.27 shows the performance of the DS UWB, while Figure 3.28 shows the performance of the TH system. There is a substantial difference between synchronous (solid) and asynchronous (dushed) performance for the DS system. The synchronous DS system has the pulses for all users transmitted at the same time. The only feature which can be used to suppress multiple access interference is the de-spreading (pulse combining) process in the receiver. When the asynchronous system is used, the lower cross correlation values that occur at different pulse alignments means that there is substantially less interference per bit to be suppressed.

The performance of the synchronous and asynchronous systems for the TH system is very similar. This is because each user has a different pulse transmit instant associated with their PR sequence, so the pulses are offset even if the time hopping frames are aligned.

The performance of the TH and DS asynchronous systems are very similar. This is to be expected in an AWGN channel with low duty cycle pulses.

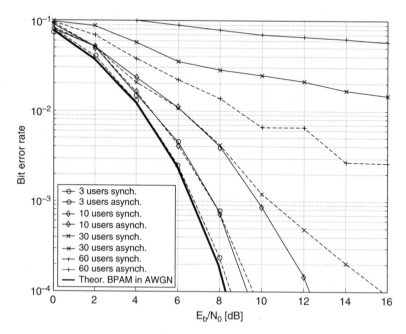

Figure 3.27   DS-BPAM in AWGN channel $T_p = 0.5$ ns

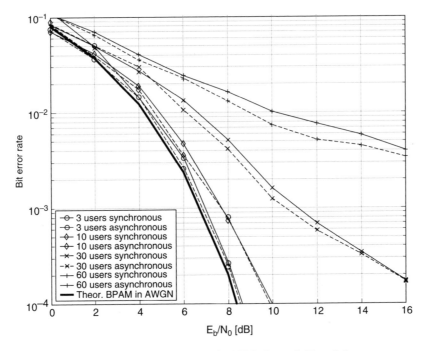

Figure 3.28   TH-BPAM in AWGN channel $T_p = 0.5\,\text{ns}$

3.6   Conclusions

UWB systems may be primarily divided into impulse radio (IR) systems and multiband systems. Multiband systems offer the advantage of potentially efficient utilization of spectrum.

However, IR systems have the significant advantage of simplicity, and so are potentially lower cost. In addition, IR is essentially a baseband technique. The IR UWB concepts investigated support many modulation schemes including orthogonal and antipodal schemes. Basic modulation must also include some form of spectrum randomization techniques to limit the interference caused by the transmitted pulse train. Both TH and DS randomization techniques were examined. Which modulation scheme to use depends on the expected operating conditions and the desired system complexity.

4

Receiver Structures

Matti Hämäläinen, Jari Iinatti, Raffaello Tesi, Simone Soderi,
Alberto Rabbachin

4.1 Introduction

UWB systems can be characterized as an extension of traditional spread spectrum (SS) systems. One of the major differences between UWB systems and traditional SS systems is the radio channel which they use. As discussed in Chapter 2, the UWB channel is extremely multipath rich. The multipath components that are combined increase the total signal power. The multipath components that are not combined lead to interference. The significantly greater number of resolvable multipath components associated with UWB system's much greater bandwidth means that many more receiver elements need to be considered. This chapter examines some of the more popular receiver structures for UWB systems, particularly the rake and modifications of the rake.

Synchronization is one of the main problems in telecommunications, navigation and radar applications. Different synchronization levels operate for carrier, code, symbol, word, frame and network synchronization. As in SS systems, code synchronization should be performed in UWB systems. When the receiver is synchronized, the received spreading code and reference code are aligned with the same phase. All of these levels of synchronization can be split into two phases—acquisition (coarse synchronization, initial synchronization) and tracking (fine synchronization). It is only possible to communicate using spread-spectrum systems if all of the necessary synchronization levels are performed with sufficient accuracy. This chapter examines code synchronization for DS and TH UWB systems.

UWB Theory and Applications   Edited by I. Oppermann, M. Hämäläinen and J. Iinatti
© 2004 John Wiley & Sons, Ltd   ISBN: 0-470-86917-8

4.2 Rake Receiver

The received signal energy can be improved in a multipath fading channel by utilizing a diversity technique, such as the rake receiver (Proakis, 1995). Rake receivers combine different signal components that have propagated through the channel by different paths. This can be characterized as a type of time diversity. The combination of different signal components will increase the signal-to-noise ratio (SNR), which will improve link performance. We will consider three main types of rake receivers.

4.2.1 Rake Receiver Types

The ideal rake receiver structure captures all of the received signal power by having a number of fingers equal to the number of multipath components. The so called ideal-rake or all rake (I-rake and A-rake) is such a receiver. (Win *et al.*, 1999; Win and Scholtz, 2003). The problem with this approach is the need for an infinite number of rake branches, which also means an infinite number of correlators. Consequently, implementation of the A-rake is not possible. The performance close to the performance in AWGN channel can be met by using the maximum ratio combining (MRC). MRC involves coherently combining all of the signal components to achieve optimal perform-ance (Proakis, 1995). Some channel estimation needs to be used to obtain channel information. The delay resolution of the receiver depends on the signal bandwidth. A different number of distinguishable propagation paths can be separated by the receiver based on the channel estimate. Figure 4.1 illustrates the concept behind the A-rake including the envelope of the channel delay profile and the propagation paths that can be separated. All of the paths that arrive within the receiver's time resolution will be seen as a single path. The energy of the single path is a combination of the energy of all of the undistinguishable paths.

A practical rake receiver implementation is a selective rake, S-rake. The S-rake only uses the L_r strongest propagation paths. Information on the channel impulse response is required in order to use the S-rake. Channel estimation algorithms must be used to obtain this *a-priori* information. The SNR is maximized when the strongest paths are detected. The link performance will be improved relative to the single path receiver. Figure 4.2 shows the multipath components used by the S-rake. The channel profile equals to the one presented in Figure 4.1. The complexity of the S-rake receiver is greatly reduced relative to the A-rake by only selecting those multipath components that have significant magnitude. In the example channel profile shown in Figure 4.2, only $L_r = 7$ of the strongest paths are selected. Depend-ing on the channel delay profile, the selected paths can be consecutive, or spread over the profile.

The partial-rake receiver, P-rake, is a simplified approximation to the S-rake. The P-rake involves combining the L_r first propagation paths. The principle behind this approach is that the first multipath components will typically be the strongest and contain the most of the received signal power. The disadvantage is that the multipath components that the P-rake receiver combines are not necessarily the strongest multi-path components, so optimum performance will not be achieved. Figure 4.3 shows the P-rake for $L_r = 7$ branches using the same channel profile as earlier examples. The

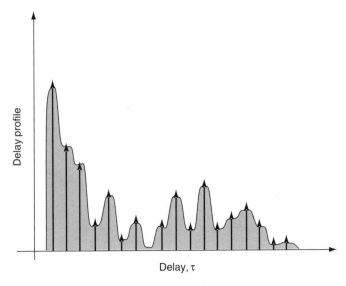

Figure 4.1   Principle of the all-rake receiver

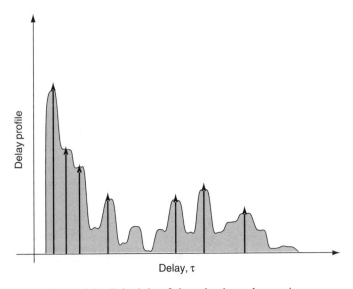

Figure 4.2   Principle of the selective-rake receiver

figure indicates that there are stronger multipath components at later delays than those which have been combined by the P-rake receiver.

Under ideal conditions, the A-rake outperforms the S-rake, which typically outperforms the P-rake. However, if the strongest propagation paths are at the beginning of the channel impulse response, the S-rake and P-rake will give the same performance.

Figure 4.4 presents the bit-error-rates for different rake receivers in modified Saleh–Valenzuela channel 2 (SV2). The number of rake fingers used, L_r, is 1, 4, 10 and 20

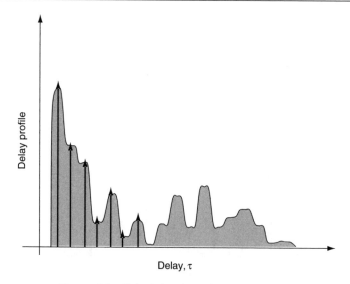

Figure 4.3   Principle of partial rake receiver

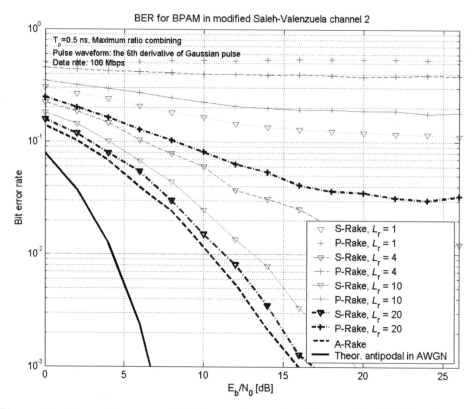

Figure 4.4   The performance of different rake receivers in modified Saleh–Valenzuela channel 2. Number of fingers $L_r = 1, 4, 10$ and 20

for both the P-rake and S-rake. The BER results for A-rake are also presented as a reference. A significant difference can be seen between the P-rake and S-rake performance because the SV2 is a non-line-of-sight model.

Figure 4.5 presents the average delay profile used in the simulations. The energy collected by the P-rake is tens of decibels less than the energy collected by the S-rake, which uses the strongest paths. The single-path S-rake uses the strongest path, so it has significantly better performance than the single-path P-rake. The system performance of A-rake is not as good as the ideal system performance in the AWGN channel because of the limited delay resolution of the A-rake receiver.

4.2.2 Detection Techniques

The rake-based receiver algorithms can be implemented in different ways, depending on the knowledge of the gain of the channel taps, in both amplitude and phase. These schemes are characterized as coherent if the phase of the channel tap is recovered, or non-coherent if the phase of the channel tap is not recovered. Channel tap amplitude recovery can be considered as a possible choice for non-coherent schemes.

Equal gain combining (EGC) is a coherent detection scheme. EGC requires a perfect estimation of the phase of each of the channel taps in order to correct the offset at the

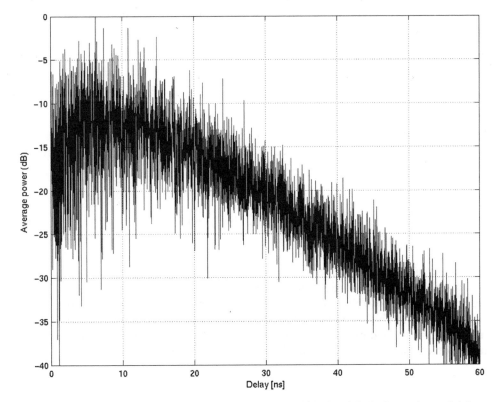

Figure 4.5 Average power delay profile of the modified Saleh–Valenzuela model 2

received signal before the detection block. The received signal for a single databit may be defined as follows

$$r(t) = \sum_{n=1}^{L_r} a_n s(t - \tau_n) + n(t), \tag{4.1}$$

where L_r is the number of recovered paths, $s(t)$ the transmitted signal, $n(t)$ the Gaussian noise in the channel, $a_n = |a_n|\, e^{j\theta_n}$ the gain, τ_n the delay of the n th multipath. θ_n will only assume values 0 and π as in the modified SV model (See Chapter 2).

Thus, the EGC decision variables will be

$$U_i^{(\text{EGC})} = \sum_{n=1}^{L_r} e^{-j\theta_n} \int_0^{T_d} r(t - \tau_n) w_i(t)\, dt \tag{4.2}$$

where T_d is the databit length and $w_i(t)$ the pulse waveform (e.g., train of pulses) representing databit 'i' at the receiver.

A second coherent scheme is maximal ratio combining. MRC involves phase recovery and estimating the received power level for each multipath. The decision variable will then assume the following form

$$U_i^{(\text{MRC})} = \sum_{n=1}^{L_r} a_n^* \int_0^{T_d} r(t - \tau_n) w_i(t)\, dt. \tag{4.3}$$

An absolute combiner (AC) can be implemented in non-coherent detection by adding the absolute values of the outputs of all the matched filters before feeding the detector. In this case the decision variables will be

$$U_i^{(AC)} = \sum_{n=1}^{L_r} \left| \int_0^{T_d} r(t - \tau_n) w_i(t)\, dt \right|. \tag{4.4}$$

The implementation of the S-rake receiver is based on the non-coherent power estimation (PE) of each single channel path. The knowledge of the channel amplitude can be used to improve the performance of the system, by weighting the output of each single correlator. This is similar to the MRC approach, except that it does not require knowledge of the channel phase. This assumption leads to a more sophisticated implementation of AC, analytically defined as

$$U_i^{(AC+PE)} = \sum_{n=1}^{L_r} \left(|a_n| \cdot \left| \int_0^{T_d} r(t - \tau_n) w_i(t)\, dt \right| \right). \tag{4.5}$$

For more information upon combining schemes, refer to Proakis (1995).

The simulations results presented below are based on a UWB system having $PG = 20\,dB$. The length of the pulse waveforms is 0.5 ns, leading to a data rate of 20 Mbit/s. In the TH case, the pulse integration gain PG_1 is 10 dB, that is, each data bit is composed of 10 pulse waveforms, whose time location is defined by the PR code.

The binary modulation schemes are PPM (for TH only), PSM and PAM (for both TH and DS). The combining approach is one of the techniques depicted above: MRC and EGC for coherent detection, AC and AC + PE for non-coherent. PE is assumed perfect in all cases. The total average signal-to-noise ratio at the receiver, E_b/N_0, is fixed to two values, 8 dB and 15 dB, respectively. However, the signal-to-noise ratio in the decision variable is less, since not all the paths have been combined.

The detection block is defined by a selective chip-spaced receiver, with a time resolution of 0.5 ns, that is, equal to the length of the pulse waveform. The channel models were simulated using at least 100 channel realizations for each case (number of fingers) examined. The channel power has been normalized to 1 over all the channel realizations used for each simulation point.

Figures 4.6 and 4.7 show the behaviour of some of the results of the simulations, which evaluated the BER of the system as a function of the number of rake fingers D for

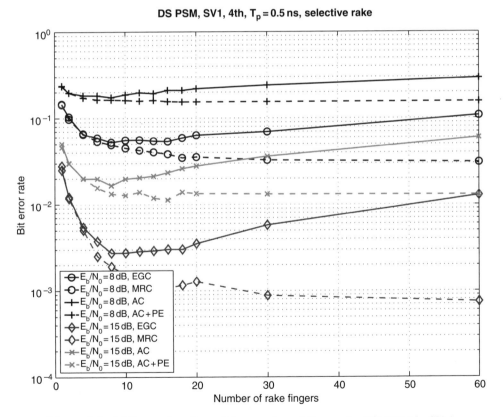

Figure 4.6 BER as a function of the number of fingers for DS-PSM in SV-1

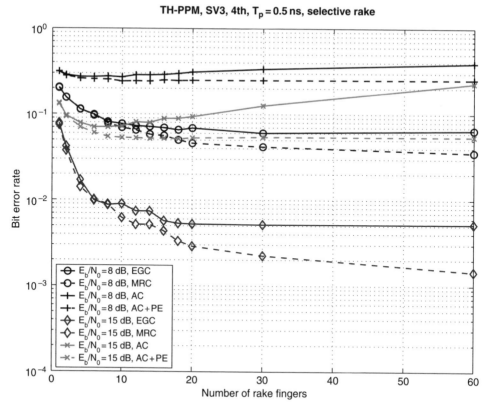

Figure 4.7 BER as function of the number of fingers for TH-PPM in SV-3

$E_b/N_0 = 8$ and $15\,dB$. In particular, Figure 4.6 represents DS-PSM for SV-1, and Figure 4.7 is TH-PPM in SV-3 case. The absence of power estimation in both the coherent approaches (EGC) and non-coherent approaches (AC) generates a minimum value of BER for a defined D_{opt}, which can be clearly chosen as the minimum. This behaviour is more evident in SV-1, where the presence of a LOS component makes the minimum BER appear for a lower D, relative to SV-3. In the MRC case, the performances of the systems are continuously improving as D increases due to the perfect weight used in estimation. Thus, the optimal value has been chosen where the BER performance tends to saturate. The improvement given by the use of coherent detection is characterized by a BER approximately 10-times lower than the equivalent non-coherent implementation for a fixed value of SNR and D.

Table 4.1 depicts the optimal number of fingers for each of the examined systems. The table gives some general trends, such as D_{opt} generally increasing with the E_b/N_0. However, this effect is more remarkable for non-coherent systems. D_{opt} is higher in SV3, due to absence of LOS component. D_{opt} is also clearly lower for non-coherent systems. Antipodal modulations show the same results for a fixed E_b/N_0 and channel model, both in terms of number of fingers and BER. Among orthogonal modulations, TH-PPM the one that shows the lowest values of D_{opt}, on average, despite the poorer BER results.

Table 4.1 Optimal number of rake fingers for different receiver algorithms

SV	E_b/N_0 [dB]	UWB system concept	Optimal number of fingers (D_{opt})			
			Coherent		Non-coherent	
			MRC	EGC	AC+PE	AC
1	8	DS-PSM	12	8	4	4
		TH-PSM	12	8	4	4
		TH-PPM	10	10	4	2
		DS-BPAM	12	10	–	–
		TH-BPAM	12	10	–	–
		DS-PSM	14	8	10	8
		TH-PSM	20	10	10	6
	15	TH-PPM	14	8	8	6
		DS-BPAM	15	10	–	–
		TH-BPAM	15	10	–	–
		DS-PSM	18	16	8	6
		TH-PSM	20	18	6	6
	8	TH-PPM	20	14	4	4
		DS-BPAM	15	15	–	–
3		TH-BPAM	15	20	–	–
		DS-PSM	20	16	16	12
		TH-PSM	16	16	16	12
	15	TH-PPM	20	18	10	6
		DS-BPAM	18	15	–	–
		TH-BPAM	18	20	–	–

4.3 Synchronization in UWB Systems

4.3.1 Basics

A survey of the published literature reveals a number of publications relating to acquisition in direct sequence spread spectrum systems (Katz, 2002). Assuming an AWGN channel, the input signal for the receiver in one user case is given by

$$r(t) = \sum_{l=1}^{L} A_l \cdot s^{(1)}(t - \tau_l) + n(t) \tag{4.6}$$

where A_l is amplitude of radio channel path l, τ_l is delay of radio channel path l, L is the number of resolvable multipath components, $n(t)$ is additive white Gaussian noise.

The delay can be presented as a portion of T_c as $\tau_l = \xi T$. The signal waveform $s(t)$ is dependent on the system concept, that is, TH or DS. In both cases, the signal waveform includes the effect of the spreading code.

The synchronization stage must provide an estimate ξ_e for the timing offset ξ (i.e. τ_e). The maximum likelihood (ML) algorithm generates several values ξ_e to evaluate when conditional probability density function of the received signal $p(r|\tau)$ achieves the maximum value (Iinatti and Latva-aho, 2001). So,

$$\hat{\tau}_e = \arg\max_{\tau_e} p(r|\tau_e). \tag{4.7}$$

As we assume white Gaussian noise channel, the conditional probability density function is readily evaluated. For the optimal estimation τ_e, we have to maximize

$$\lambda(\tau_e) = \int_0^{T_o} r(t)s(t, \tau_e) \, dt$$

$$= \int_0^{T_o} A_l \sum_{k=0}^{\infty} \sum_{j=0}^{N-1} w(t - kT_b - jT_f - \tau). \tag{4.8}$$

$$\sum_{k=0}^{\infty} \sum_{j=0}^{N-1} w(t - kT_b - jT_f - \tau_e) \, dt + n_c(t)$$

where $n_c(t)$ is the additive noise component, and the information signal is in the form of TH-UWB as described in Chapter 3. In practice, the algorithm is too complex to implement, and some approximations must be used. In the so-called serial search strategy, the phase of the local code (τ_e) is changed step-by-step in equal increments. In this way, the correct position within the uncertainty region can be found (Iinatti, 1997). Serial search acquisition can be implemented using active correlation measurement techniques or passive correlation measurement based on matched filtering (MF).

4.3.1.1 Synchronization Schemes

UWB systems use very large bandwidths, and therefore a dense channel multipath profile where many components can be distinguished from the received signal. The multipath channel then introduces more than one correct synchronization cell. From the perspective of code synchronization, this phenomenon causes problems: the energy of the signal is spread over many multipath components, and the energy of each path is very low. Therefore, the paths are difficult to acquire. Depending on the receiver structure (type and length of rake), a number of paths should be acquired.

In DS systems, the uncertainty region corresponds to a multiple of the code length. In TH systems, the uncertainty region is divided into a number of cells (Cu). The number of cells depends on the number of possible pulse positions combinations in a bit interval. Recently, an algorithm called chip-level post-detection integration (CLPDI) was proposed for code acquisition in direct sequence CDMA systems (Iinatti and Latva-aho, 2001). Algorithm is suitable for synchronization in multipath environments. Once the CLPDI algorithm finds one of the possible synchronization cells, an additional sweep has to be performed to acquire the necessary number of paths. The aim of this initial code acquisition in a multipath channel is to find a starting point to reduce the multipath search time. In this chapter, the method is applied for time hopping UWB system due to nature of UWB signals.

Figure 4.8 presents the synchronisation algorithm for a multipath environment proposed for DS CDMA (Iinatti and Latva-aho, 2001). In DS systems, the impulse response of the matched filter is the time-reversed replica of the spreading code. This means that the impulse response has coefficients given by the spreading code, and the delays between the consecutive coefficients are T_c. The MF output signal is proportional to the autocorrelation function (ACF) of the spreading code. The sampling at the output of the MF is made at least at the chip rate. The MF is followed by a threshold comparison, and if the threshold T_h is crossed, the acquisition process ends. There is detection if the threshold is crossed by the ACF at the zero delay. This occurs with probability of detection, P_d. If the threshold is crossed with some other delay, a false alarm occurs. This happens with the probability of false alarm, P_{fa}. False alarms may be catastrophic, and cause total miss of the correct code phase. Because the MF gives its peak when the code is inside the filter, a multipath channel leads to several peaks.

Chip-level post-detection integration performs post-detection integration at the chip level, i.e., a number of consecutive samples at the output of the MF are combined. CLPDI is performed over m samples and CLPDI output is sampled at multiples of mT_c as is presented in Figure 4.8. Because the sampling is done as multiples of mT_c, consecutive samples at the output of CLPDI are uncorrelated. In addition, the uncertainty region, i.e., the number of cells to be tested in the acquisition, is decreased relative to the pure MF acquisition. The number of cells in the uncertainty region is now reduced to $C_m = Cu/m$. In addition, the number of potentially correct cells to be tested is reduced, as can be seen by comparing Figures 4.9 and 4.10.

In time hopping systems, the MF collects the samples together according to delays between consecutive chips (pulses as presented in Chapter 3), i.e., the spreading code sets the delays between consecutive '1's in the impulse response. Therefore, the MF waits until the chips (pulses) arrive in predetermined time slots inside frames.

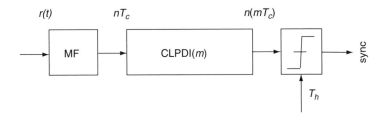

Figure 4.8 MF with CLPDI

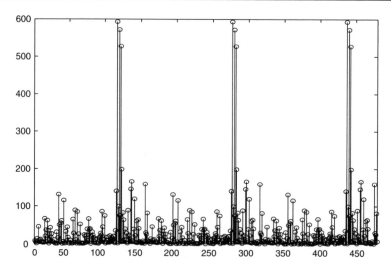

Figure 4.9 MF output without noise as function of time in channel model 1, number of pulses in bit is $N = 10$ and number of pulse position in frame is $M = 16$

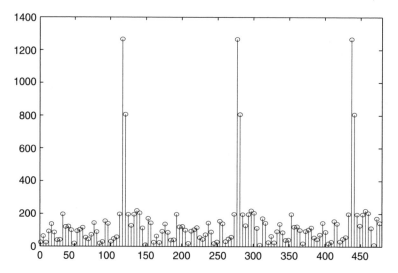

Figure 4.10 CLPDI output without noise as function of time in channel model 1, $N = 10$, $M = 16$ and $m = 4$

4.3.2 Performance Measures

The specifications of satisfactory performance measures for synchronization will depend on the particular application. The main issue is the time that elapses between the time the synchronization starts and the time of acquisition. There are two basic scenarios—when there is an absolute time limit and when there is no absolute time limit. If there is no time limit, the most interesting parameter is the mean acquisition time, $T_{MA} = E\{T_{acq}\}$ and possibly the variance of the acquisition time $\sigma^2_{T_{acq}}$. This is the case when a data or pilot signal is always present, that is, when the link operates continuously (Polydoros, 1982). The mean acquisition time is defined as the expected value of the time that elapses between the initiation of acquisition and the completion of acquisition. If there is a time limit, the parameters are the acquisition time T_{acq} and the time limit T_s, If there is a time limit, a better performance measure is the probability of prompt acquisition $\Pr\{T_{acq} \leq T_s\}$, which is called overall probability of detection P_d^{ov}. The complement of the probability of prompt acquisition is the overall probability of missing the code $P_m^{ov} = 1 - P_d^{ov}$. Missing the code occurs in systems where data transmission starts after a certain time interval T_s from the initial system start up, i.e., burst communication (Polydoros, 1982).

The acquisition time, and therefore also T_{MA} and P_d^{ov}, are functions of several parameters. The most important of these are the probability of detection P_d and the probability of false alarm P_{fa}. The probability of detection is the probability that the decision is the 'correct cell' when the correct cell of the uncertainty region is being tested. The probability of false alarm is the probability that the decision is the 'correct cell' when a false cell is being tested. If the decision is based on threshold comparison, threshold T_h plays an important role concerning T_{acq} because both P_d and P_{fa} are functions of this threshold. The other parameters are the time spent evaluating one decision variable T_i, the penalty time T_{fa}, and the number of cells in the uncertainty region Cu.

In this context, mean acquisition time is used as a performance measure. It can be calculated for the DS CLPDI case as (Iinatti and Latva-aho, 2001):

$$T_{MA} = \frac{P_M^{L/m}[LT_c + (N-L)(T_c + T_{fa}P_{fa})]}{1 - P_M^{L/m}}$$

$$+ \frac{[NT_c + (N-L)(T_c + T_{fa}P_{fa})]\sum_{i=0}^{L/m-1} iP_M^i}{N \sum_{i=0}^{L/m-1} P_M^i/m} + mT_c \qquad (4.9)$$

$$+ \frac{(N-L)(N-L+m)(T_c + T_{fa}P_{fa})}{2N} + NT_c$$

where the uncertainty region N is the length of the code, no a-*priori* information of the correct code phase exists at the beginning of acquisition, P_{fa} is the probability of false alarm at the output of the CLPDI, $P_M = 1 - P_d$ where P_d is the probability of detection in the correct code phase, $T_{fa} = K_p T_c$ is the penalty time caused by a false alarm, T_c is chip interval.

In a multipath propagation environment without the CLPDI block, the T_{MA} is evaluated with a similar equation (Ramirez-Mireles and Scholtz, 1998a).

4.3.2.1   Performance of CLPDI

Figure 4.11 presents the TH UWB simulation model with matched filter synchronization. The figure also describes the outputs from different parts of the system.

Chip correlation (point A in Figure 4.11) is performed by correlating the received signal with the template waveform which has the same shape as the received signal (after the antenna). The pulse waveform in the transmitter in the simulations is a Gaussian pulse. Therefore, the reference pulse in the receiver is the second derivative of the Gaussian pulse because of the effect of the transmitting and receiving antenna (Ramirez-Mireles and Scholtz, 1998a). Only one sample per T_c is used so as to reduce the complexity in the simulations. This can be done when the pulse width (T_p) is smaller than chip length (T_c), and when chip synchronism is assumed. The output of the chip correlator is a sequence that contains the information about the pulse locations within the received sequence in every frame. The chip correlator output is fed into a code matched filter (CMF). The filter is matched to the whole spreading code (point B in Figure 4.11). In the TH system this means, that delays between consecutive '1's in impulse response is not fixed, but instead depends on the code phase (as seen in Figure 4.12. After the filter, a threshold comparison is used (point C in Figure 4.11). Figure 4.12 presents an example of the output of code matched filter, as well as describing the correct synchronization points.

In the CLPDI simulations, the CLPDI block is inserted after the code matched filter in the block diagram of the TH-UWB synchronizer (Figure 4.11).

4.3.2.2   AWGN Channel Performance

We will first consider a channel with static multipath. For a static channel, the equal power path case represents the worst scenario (Iinatti and Latva-aho, 2001). One sample per chip is used to achieve lower simulation times, and chip synchronism is assumed. We use a code length $N = 80$ and penalty time $T_{fa} = 100 T_b$. A constant false alarm rate

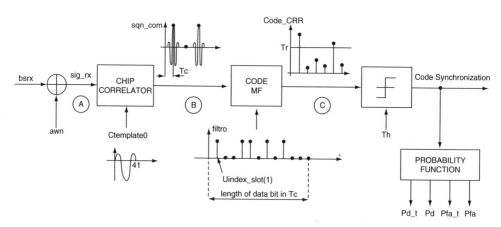

Figure 4.11   Simulation model for TH-UWB synchronizer in AWGN channel

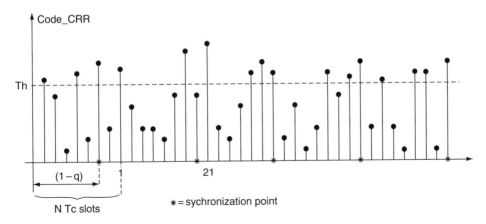

Figure 4.12 Output of the code matched filter when six bits are transmitted

criterion (CFAR) is used when setting the threshold of the comparator ($P_{fa} = 10^{-2}$). The T_h was found before simulations, so that the required P_{fa} was obtained for each SNR value. Simulations were performed over 100 bits, i.e., the number of the correct code phases is $100L$ and the number of false code phases is $100(80 - L)$.

Figure 4.13 presents T_{MA} performance of the synchronizer with and without CLPDI for two channels, and Figure 4.14 presents T_{MA} performance of the synchronizer with and without CLPDI for four-path channels. In a four-path channel, m in CLPDI

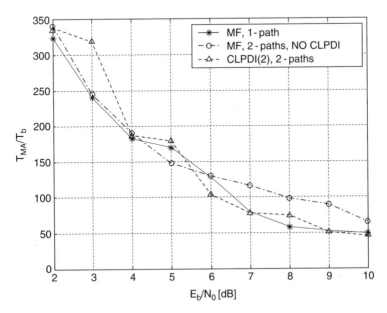

Figure 4.13 T_{MA} with and without CLPDI, $N = 80$, $L = 2$, $T_b = NT_c$, $T_{fa} = 100$, $P_{fa} = 10^{-2}$

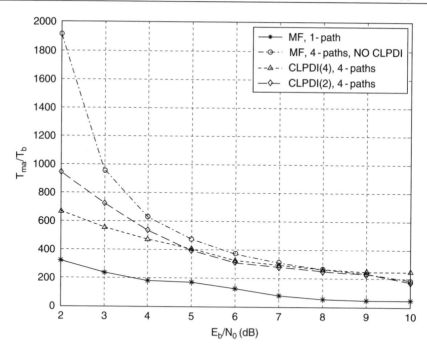

Figure 4.14 T_{MA} with and without CLPDI, $N = 80$, $L = 4$, $T_b = NT_c$, $T_{fa} = 100$, $P_{fa} = 10^{-2}$

is either 2 or 4, P_d and P_{fa} are obtained by simulation, and T_{MA} is calculated from equation (4.9) utilizing the uncertainty region of TH system. T_{MA} is also presented for a one-path channel. In TH-UWB systems, CLPDI is an improvement over conventional MF synchronizers because there are fewer total cells to be examined, and because of the increased P_d in correct cells. In a four-path channel, performance is improved by using both m values 2 and 4. The behaviour is similar to WCDMA systems (Iinatti and Latva-aho, 2001).

4.3.2.3 Performance in Saleh–Valenzuela Channels

There is a large number of resolvable paths in UWB systems, more than 100, because of the fine path resolution of the UWB channel. The following simulation results only consider the most powerful paths for the code acquisition. The number of cells or chip positions on which the receiver can be considered synchronized is defined by the 20 strongest paths in the MF case. When m-CLPDI is used, we need to make sure that the 20 synchronization positions defined in the MF case are included in the synchronization cells for the m-CLPDI algorithm. Since it is possible that the 20 synchronization cells are non consecutive, the number of synchronization cells in the m-CLPDI case can be bigger than $20/m$. Once one of the synchronization cells is found, a new acquisition process starts. Verification extra-time is added for each false alarm event occurred during the acquisition process.

The simulations consider four modified Saleh–Valenzuela channel models presented in Chapter 2. The pulse width T_p is fixed to 1 ns. The code length is 160, the number of frames per bit is $N = 10$, the number of possible pulse position in each frame is $M = 16$, the penalty time in the case of false alarm is $T_{fa} = 100T_b$ and the required $P_{fa} = 0.01$. The synchronization process is performed over a data packet of 30 bits and the channel is static over the each packet. A new data block is transmitted every time the synchronization is reached or the data packet has been explored without synchronization.

Figures 4.15 to 4.18 present the mean acquisition time of the code MF for $m = 4$, 10 and 20 and the CLPDI block versus SNR for each of the Saleh–Valenzuela channel models. SNR is defined as the E_b/N_0 at the receiver side. E_b is the collective bit energy spread over all the paths in the channel. Using CLPDI, the reduced code acquisition time is evident for all the channel models. The mean acquisition time increases in absolute time from both MF and CLPDI passing from the CM1 to the CM4. The reason is the reduced energy borne by the 20 strongest paths whilst moving from CM1 towards CM4. Otherwise, passing from the CM1 to the CM4 increases CLPDI's performance improvement relative to the MF case. As the channel moves from the CM1 to the CM4, the channel gets closer to the situation of having paths with equal average power that gives the best result, as has been shown for data detection using diversity techniques (Proakis, 1995). In CM1, the extra gain achieved using $m = 20$ is not remarkable. This is because of the reduced CM1 delay spread relative to the other channel models.

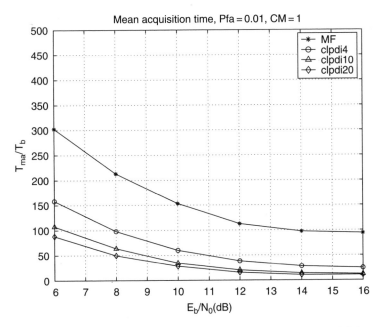

Figure 4.15 T_{ma}/T_b as a function of E_b/N_0 in channel model 1

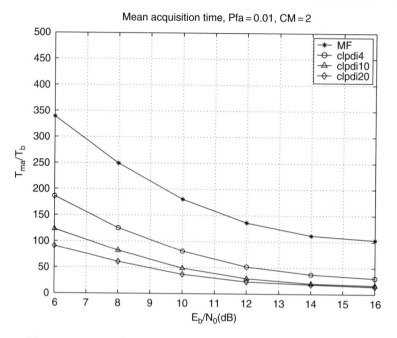

Figure 4.16 T_{ma}/T_b as a function of E_b/N_0 in channel model 2

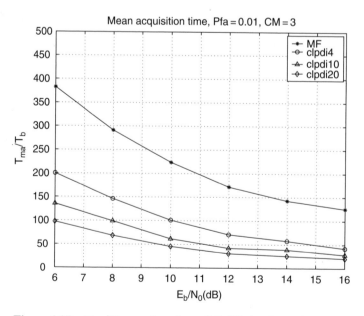

Figure 4.17 T_{ma}/T_b as a function of E_b/N_0 in channel model 3

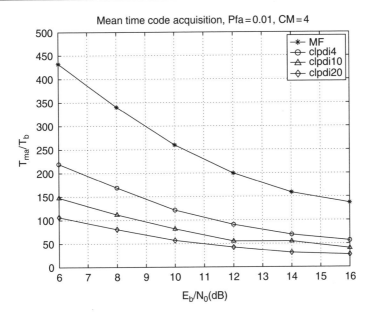

Figure 4.18 T_{ma}/T_b as a function E_b/N_0 in channel model 4

4.4 Conclusions

This chapter examined the concepts behind some of the most common UWB receivers. The chapter described the A-rake, S-rake and P-rake and analysed their relative performance. The rake combines the many multipath components in the multipath-rich channel. The complexity of the A-rake makes it impractical for many applications. The S-rake and P-rake combine only a subset of the resolvable multipath components. Whilst being much less complex, the S-rake and P-rake are not able to combine all the received signal energy and so have substantially worse performance than the A-rake. The S-rake requires channel estimation to select the most significant multipath components. This added complexity greatly improves the performance compared with the P-rake, especially in non-LOS channels.

This chapter also applied a chip-level post-detection integration code synchronization method to a time-hopping UWB system. The method was used in order to utilize dense multipath propagation in a UWB environment during the code acquisition process. The method increases the probability of detection of the burst of multipath components, i.e., it decreases the mean acquisition time for finding the existence of multipath profile. The simulation results indicate that the method improves the acquisition performance, and therefore is usable in a UWB communication system.

5

Integrated Circuit Topologies

Sakari Tiuraniemi, Ian Oppermann

5.1 Introduction

UWB is a spread spectrum technology. However, UWB differs from conventional spread spectrum technologies in that UWB transmits information through short pulses or a 'chirped' signal rather than transmitting information on a modulated continuous carrier signal. The information is typically superimposed by a pulse modulation method. The bandwidth occupied by an UWB system is far greater than in traditional spread spectrum systems. A UWB data communication system is often referred to as impulse radio (IR). Typically, impulse radio utilizes time hopping pulse position modulation (TH-PPM) scheme (see Appendix 1).

UWB systems have some significant advantages over conventional spread spectrum systems. First, the transceiver has a relatively simple structure as some of the functional parts that increase the complexity of traditional radio systems are not necessary. This reduces the required human resources and financial investment. Second, the low transmission power means that less power is consumed. Third, the large bandwidth makes the detection of the signal quite difficult for unintended receivers. Fourth, the achieved bit rate of more than a few hundreds of Mbps is significantly more than conventional spread spectrum systems.

In this chapter, some of the architectures used in UWB data communication systems are introduced and investigated, based on the information set out in the current available literature.

In particular, we will examine the pulse generator, a critical part of the transmitter, which generates the transmitted waveform. The transmitted waveform must satisfy the frequency mask defined by the Federal Communications Commission (FCC) (Federal

UWB Theory and Applications   Edited by I. Oppermann, M. Hämäläinen and J. Iinatti
© 2004 John Wiley & Sons, Ltd   ISBN: 0-470-86917-8

Communications Commission, 2002b) or satisfy other applicable local radiation regulations. One of the most common waveforms is the first derivative of the Gaussian pulse, also referred to as the monocycle.

We will also examine the correlation-based receiver. A conventional SNR maximizing based correlator receiver requires a template waveform that exactly matches the received waveform. Such a distorted and delayed waveform is difficult to generate. The multiplier and integrator of the correlating circuit must be extremely fast to process the short time domain pulses. UWB receiver structures often ignore or coarsely approximate the pulse shape, which significantly reduces the complexity and required speed of operation, even though it also reduces the received SNR. This approach seems to be quite popular amongst current receiver designs (for example, see Lang, 2003) and is a recommended approach.

5.2 Ultra Wideband Basic Architectures

Impulse radio is a UWB digital data communication system for low power, low range, applications typically utilizing TH-PPM. At the block level, the transmitter is very simple. It consists of a pulse generator and a digital timing circuit that controls the timing of transmission. These blocks are presented in Figure 5.1.

The block marked by τ is the timing circuit that is responsible for PPM and PN coding. It provides a timing signal, or a trigger, for the pulse generator. In some presentations the timing circuit is replaced by a programmable delay (Withington, 2004).

The clock oscillator determines the pulse repetition frequency, PRF, of the system. It can be either a crystal oscillator or a custom designed oscillator, the first being the most feasible in high frequency systems such as impulse radio. The pulse generator produces the desired waveform. The pulse generation can be realized in a number of ways.

One of the great benefits of a UWB transmitter relative to continuous-wave transmitters is that there is no need for complex circuits such as the power amplifier (depending on the application) and frequency synthesizer, which contain circuits such as the phase-locked loop (PLL), voltage controlled oscillator (VCO) and mixers (Taylor, 1995; Foerster *et al.*, 2001). These are the most complex components of conventional transmitters, and these components make conventional transmitters relatively difficult and expensive to design and implement. In contrast, an UWB transmitter is relatively

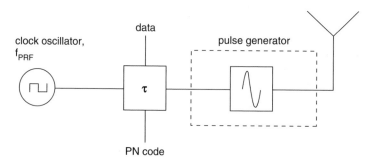

Figure 5.1 Transmitter top level schematic

inexpensive and moderately easy to design and implement because the UWB transmitter does not have these components.

The optimal receiver for signals transmitted over an AWGN channel is a correlation or a matched filter receiver, since it maximizes the SNR (Proakis, 1995). The receiver examined in this chapter consists of a low-noise amplifier (LNA), a correlation circuit and a circuit to provide the template waveform for the correlation. These blocks are presented in Figure 5.2.

After the received signal is amplified, it is correlated with the template waveform. The output of the correlation is processed by a bit decision circuit that decides which bits are carried by the received pulses. The baseband signal processing circuit is responsible for the bit decision.

To maximize the processing gain and SNR, the template waveform should be the same as that of the received signal. Such a signal is difficult to generate in practice, since the transmitted pulse is distorted by the antennas and the channel. The need to make the template waveform the same as that of the received signal also makes the receiving circuit more complex. One way to avoid this complexity is to approximate the template waveform by using the transmitted pulse, or to make very coarse approximations such as a rectangular pulse. It is also possible to ignore the template waveform altogether and rely on the pulse shaping caused by the finite bandwidth of the transmitting and receiving antennas (Lang, 2003).

In an ideal environment, the received pulse shape is the second derivative of the signal transmitted (third derivative of the Gaussian pulse) since both antennas act as differentiators. Figure 5.3 shows the autocorrelation functions of the first and third derivatives of the Gaussian pulse as well as the cross-correlation between the first and third Gaussian pulses. The amplitude of the correlation result between the first and the third derivatives is 80 % of the autocorrelation of the received waveform. This means that using the transmitted waveform as the template reduces the performance of the correlator by less than 1 dB.

The correlation circuit consists of an integrator, and a multiplier that multiplies the received signal with the template waveform. The result of the multiplier is integrated

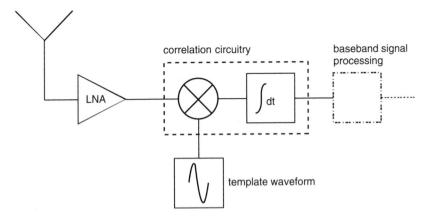

Figure 5.2   Receiver top level schematic

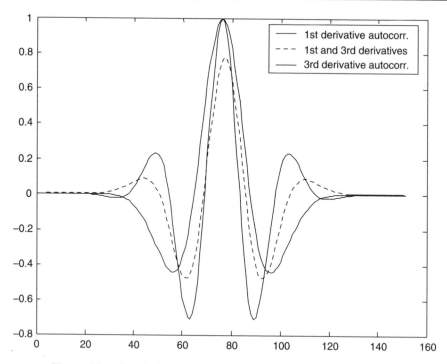

Figure 5.3   Correlation results of transmitted and received signals

over the bit duration to maximize the received signal power and to minimize the noise component. Having a train of pulses to integrate over, the correlated signal is raised from the noise and the possible signals of other users. From this it can be seen that the more pulses there are in a pulse train, the better SNR is attained, since more correlated energy is put into each symbol.

In addition to providing an accurate synchronization, performing the correlation at the required speed is probably the biggest issue. Both the multiplier and integrator must be fast enough to process each pulse.

5.3   Review of Existing UWB Technologies

This section reports some of the architectures used in UWB communication applications, with a particular focus on the pulse or waveform generation and its implementation. This section also considers the used data modulation scheme and the correlation receiver. These topics, combined with the transceiver architecture and antennas, form the basis for the communication system.

The information is gathered from patents and the publications listed in the references. The architectures are chosen to give examples of the possible structures used in UWB communications, with a particular focus on the pulse or waveform generation and implementation.

5.3.1 Time Domain Corporation: PulsOn Technology

Time Domain Corporation's PulsOn technology (Time Domain Corporation, 2003) is presented in Figure 5.4. This, or a variant of it, is the most common architecture used in UWB communication systems. A pulse generator generates the waveform (e.g. a Gaussian monocycle), and the waveform is then provided to the transmitting antenna. The pulse transmitting time is controlled by a programmable time delay (PTD), which uses the signal coming from the clock oscillator to create a timing signal for the pulse generator by means of pulse position modulation. The programmable time delay gets its control signal from the modulator and code generator. The modulator modulates the incoming data using PPM or an additional modulation scheme, and the code generator gives an individual pseudo-random (PR) code for the modulated data. Power amplification is not required because of the low transmit power (below the noise level).

The receiver is based on the correlation technique. The correlation technique is the optimal technique for a signal of such power since it maximizes the SNR (Proakis, 1995). The received signal is multiplied with a template waveform generated in the receiver, and the result is then integrated over several periods of the received pulse train. The correlator converts the received signal directly into a baseband signal, which may then be further processed using baseband techniques to improve performance. The signal processor also provides acquisition and tracking control to the PTD.

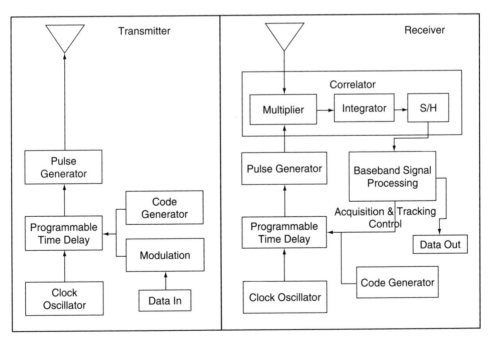

Figure 5.4 PulsOn transceiver top level schematic (From Time Domain Corporation website)

5.3.2   Time Domain Corporation: Sub-Carrier Technology

The Time Domain Corporation patent also includes another architecture developed by the company (Time Domain Corporation, 2003) (see Figure 5.5). The architecture's basic structure is the same as Time Domain Corporation's original PulsOn technology, except that the second architecture utilizes sub-carrier technology to realize an additional modulation scheme.

The time base in Figure 5.5 provides a clock oscillator, which again functions as the PRF source. The code time modulator and code source create a coded timing signal from the PRF source. In addition, the information source is fed to a sub-carrier generator and modulator block, which creates a sub-carrier signal (i.e. a piece of sine wave) and modulates it with the information signal to create a modulated sub-carrier signal. A sub-carrier time modulator then mixes the resultant modulated sub-carrier signal with the previously created coded timing signal from the PRF source, and provides the result to an output stage as its time controlling signal.

This technique provides further possibilities for channel coding and signal modulation. For example, by using different sub-carrier frequencies, more channels can be utilized or different information can be transmitted simultaneously. If different sub-carriers are used, the receiver must separate the carriers from the received signal. This may be achieved by band-pass filtering for example.

5.3.3   MultiSpectral Solutions, Inc.

MultiSpectral Solutions introduces three schemes to generate pulses or bursts of pulses in their patent (Parkway, 2001). They have divided the three pulse generation schemes into two classes. The first class consists of two types of pulse generation scheme. The first pulse generation scheme, presented in Figure 5.6, utilizes an impulse generator and a mixer (switch) to chop the signal coming from an oscillator, thereby providing a train of bursts to the subsequent circuitry.

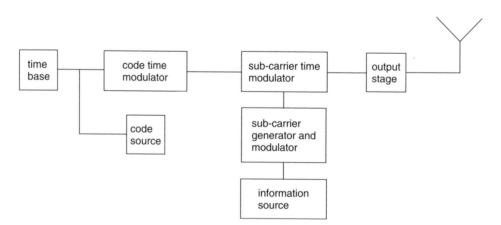

Figure 5.5   Transmitter utilizing sub carrier technology

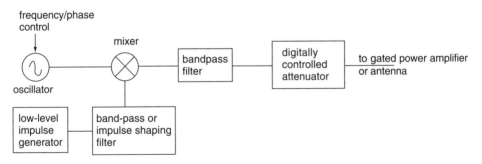

Figure 5.6   First class of transmitter

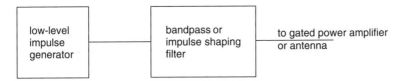

Figure 5.7   Special case of the first class of transmitter

The second pulse generator scheme in the first class is shown in Figure 5.7. A bandpass or a pulse shaping filter is directly excited by a low level impulse, so that a mixer and oscillator are not needed. This is functionally equivalent to MultiSpectral Solutions's first pulse generation technique in the special case of having a zero oscillator frequency (i.e., d.c. source).

The band-pass or pulse shaping filter (set out in Figures 5.6 and 5.7) shapes the incoming impulse, and thus provides the wanted centre frequency or bandwidth for the transmitted signal. This could also be achieved by adjusting the width of the impulse. The transceiver is less complex since there is no need to generate the derivative. The trade-off is that the circuit is relatively more expensive, since very short impulses require fast switches and digital signal processors to deal with the fast calculations.

MultiSpectral Solutions's second class consists of the third type of pulse generator scheme. It is shown in Figure 5.8. The impulse generator and filter combination in

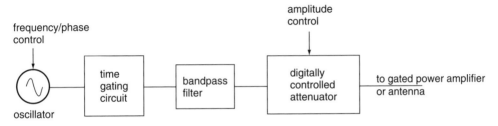

Figure 5.8   Second class of transmitter

MultiSpectral Solutions's first class of pulse generator schemes is replaced with a timing circuit. The timing circuit functions as a switch providing a short time frame for the oscillating signal to propagate through it. The response can be described as an amplitude modulation of the oscillating signal. The amplitude is defined by the ramp generated by the gating circuit.

5.3.4 XtremeSpectrum Inc.: Trinity

XtremeSpectrum's transceiver architecture is covered in a 2001 patent (McCorkle, 2001) and consists of an interface and three parts: transmitter, receiver and radio controller. The basic idea of the transceiver is the same as in the patented architectures in Sections 5.3.2. and 5.3.3 and will not be analysed further.

The waveform or pulse generation in this architecture is interesting. The idea is to produce two short pulses, which are half the length of the desired monocycle pulse. These short pulses, S1 and S2 in Figure 5.9, are then combined by a Gilbert cell, which is used as a differential mixer. As a result, the mixer produces the monocycles presented in Figure 5.9 which is redraw from (McCorkle, 2001). The polarization of the monocycle depends on the data bit A, which is the other input of the Gilbert cell. When A is low, the monocycle begins with negative amplitude, and when A is high, the monocycle begins with positive amplitude. The Gilbert cell will be introduced in a later section.

In XtremeSpectrum's patent, the Gilbert cell functions as a modulator. The short pulses are multiplied either by 1 or -1 depending on the data bit 'A'. An example of the BPAM modulation attained by the mixing function at the Gilbert cell is persented in patent (McCorkle, 2001).

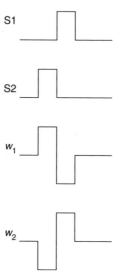

Figure 5.9 Ideal pulses and waveforms

5.3.4.1   Pulse Generation by Avalanche Transistor

In Morgan (1994), an avalanche transistor is used to generate a monocycle pulse. The pulse generation is based on operating the transistor in avalanche mode, which requires a high voltage. High-voltage solutions are not covered in this book, and therefore this method is introduced for completeness only.

The bias circuit of the pulse generator (shown as a dashed line in Figure 5.10) provides a fixed bias voltage across the capacitor C7. This voltage level is about 100–130 V, which is near the transistor's (TR3) avalanche breakdown voltage. This voltage is also provided across charging capacitor C8, which dumps a large charge through transistor TR3 when the avalanche breakdown voltage is reached. This is achieved by the pulse position modulated message signal, which is fed to the base of the avalanche transistor.

As the avalanche breakdown takes place, the emitter voltage of TR3 rises dramatically for a short time ($\sim$10 ns), and then falls to negative equivalent voltage due to the inductance connected between the emitter and earth. After the charging capacitor C8 has dumped its charge, it takes a few tens of microseconds to attain the level required for a new avalanche break down. This pulse generation method is feasible for applications with high voltage levels and monocycles with relatively large length, such as RADAR.

5.3.5   Coplanar Waveguides

A recent pulse generation method (Lee *et al.*, 2001a) is based on step recovery diode (SRD), Schottky diode and charging and discharging circuitry (see Figure 5.11). The SRD provides an impulse, which is high-pass filtered in a RC-circuit. The result is a Gaussian-like pulse, which is fed to a pair of transmission lines. The generated pulse is divided in two and propagates in both branches after the capacitor C. The first half of the pulse propagates directly to the load resistor, and the other half of the pulse propagates to the short. The transmission lines are designed to have such a length that the propagation delay of the second half of the pulse (the one propagating to the short) is equal to the length of the pulse. The pulse will be inverted when the pulse reflects from the short circuit in the end of the transmission line. The resulting pulse seen across the

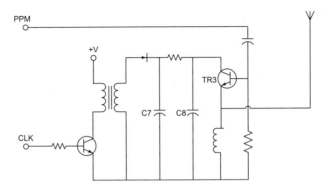

Figure 5.10   Transmitter driver including the pulse generator

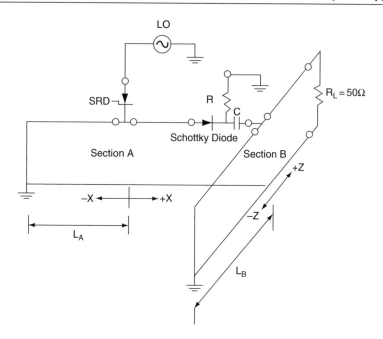

Figure 5.11   Monocycle generator

load is the superposition of the two branches. The pulse width is controlled by the SRD. The voltage across the load is shown in Figure 5.12.

Another embodiment of this development has been introduced (Lee *et al.*, 2001b). The basic idea is the same as in (Lee *et al.*, 2001a), except that there is an additional MESFET as an amplifying unit.

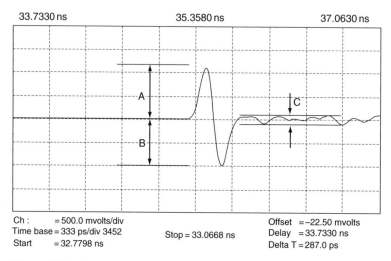

Figure 5.12   Monocycle generated by the circuit shown in Figure 5.11

5.4 Integrated Circuit Topologies

This section introduces the basic circuit topologies used to implement the transceiver on an integrated circuit (IC), and briefly introduces high-speed integrated circuit processes and designs.

The basic design flow of high-speed circuits is presented in Figure 5.13 (Häkkinen, 2002). The target specifications are set by customer or system-level requirements. The designer's task is to produce a circuit meeting these specifications after a suitable technology has been chosen.

The design process typically means iterating through the steps in Figure 5.13 several times before the customer or system specifications are met. The most common iterative operation is that of optimizing the circuit performance to meet the speed, distortion and matching specifications. This may be very challenging in portable devices because of the rigid power consumption requirements.

In radio frequencies (RF), the IC design is especially difficult if the IC process is not optimized for RF. This especially concerns CMOS which is optimized for digital circuits. In addition, the processes that are suitable for analogue design are pushed to their limits, since the maximum operating frequency is close to the signal frequencies (Häkkinen, 2002).

The circuits are strongly sensitive to component parasitic and model errors at high frequencies and near the limits of the process. This makes the designer's task even more difficult, since the designer has to rely on inaccurate device and parasitic models. The designer's expertise becomes very important.

RF circuits are also very sensitive to the layout design and the packaging that is used. Layout parasitic (such as capacitance and inductance) cause mismatch, signal crosstalk and changes in circuit response. In addition, individual circuits may disturb each other

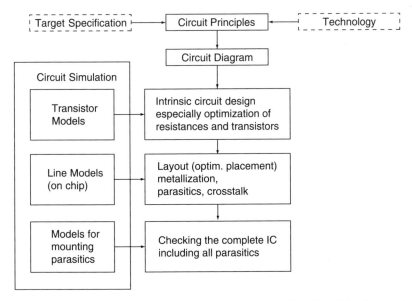

Figure 5.13 Basic steps in the design of high-speed circuits. For simplification the iteration loops between the different steps are not shown

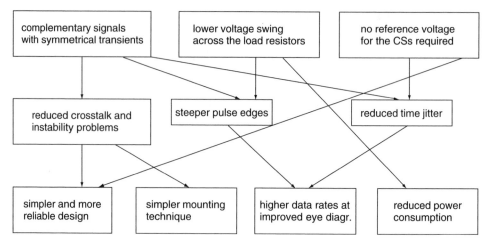

Figure 5.14 Advantages of differential operation (compared with single-ended operation)

through crosstalk, inductive cross coupling and noise induced to the internal power lines. These effects can be minimized by careful layout design and yield analysis. Again, the designer's own expertise is important. Further discussion of layout design and packaging is set out in Häkkinen (2002).

Due to these issues , the designer has to use careful design techniques and topologies. In RF circuits, it is crucial to have as little interference as possible, especially for wideband circuits, in which the interference is more likely to appear in the signal band. One widely used design practice is to use differential signals and structures. Using differential signals instead of single-ended signals minimizes the effect of a number of external and internal interfering signals, such as noise or variations of the power source. Some advantages of using differential structures are presented in Figure 5.14 (Häkkinen, 2002).

Differential circuits are required for processing differential signals. Hence all topologies examined in this chapter are built using differential pairs, except for current source and source follower. We will consider the differential-pair topologies first.

The topologies presented in this section are CMOS circuits. All these circuits may also be realized by other technologies. However, the equations presented are not necessarily applicable to non-MOS technologies. The equations for other types of active components can be found in the literature (Gray and Meyer, 1993; Johns and Martin, 1997).

5.4.1 Source Coupled Pair

The differential pair (i.e. emitter/source coupled pair or transconductance stage) is one of the basic and most common building blocks used in integrated circuits. Differential pairs can be used to process differential signals, and these circuits can be coupled to each other easily, for example, without the use of coupling capacitors. They may be used as amplifiers as well as an input stage or buffers (Gray and Meyer, 1993).

Figure 5.15 presents a source coupled pair and its input voltage versus output current curve (Babanezhad and Temes, 1985). The transistors are assumed to be identical and operating in the saturation region in all the schematics presented in this chapter.

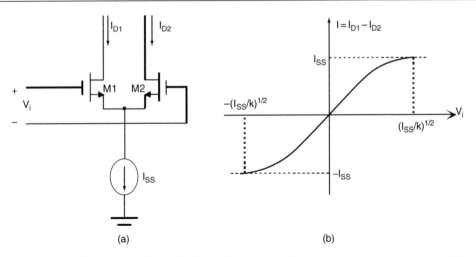

Figure 5.15 Source coupled pair (a) and its input voltage vs. output current curve (b)

The operation of the transconductance stage can be illustrated with the (low frequency) small signal model (based on the T-model, Figure 5.16) (Johns and Martin, 1997), from which the output current can be determined as a function of input voltage.

Defining the differential input voltage as $V_{in} = V_i^+ - V_i^-$, the small signal drain current is determined by (Johns and Martin, 1997)

$$i_{d1} = i_{s1} = \frac{V_{in}}{r_{s1} + r_{s2}} = \frac{V_{in}}{1/g_{m1} + 1/g_{m2}} \tag{5.1}$$

where i_{d1} is the small signal drain current, i_{s1} is the small signal current through resistor, r_{s1}, r_{s2} are the source resistances, V_{in} is the differential input voltage and g_{m1}, g_{m2} are the transconductance values of transistors M1 and M2.

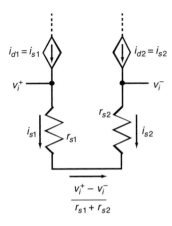

Figure 5.16 Small signal model of source coupled pair

When the two transistors, M1 and M2, are identical, the transconductance is $g_m = g_{m1} = g_{m2}$ which gives equations (Johns and Martin, 1997)

$$i_{d1} = i_{s1} = \frac{g_m}{2} V_{in} \tag{5.2}$$

$$i_{d2} = -i_{d1} = -\frac{g_m}{2} V_{in}. \tag{5.3}$$

By defining the differential output current as $i_{out} = i_{d1} - i_{d2}$, we have the following

$$i_{out} = g_m V_{in}. \tag{5.4}$$

The gain of the transconductor is defined by its transconductance g_m

$$A = \frac{i_{out}}{V_{in}} = g_m$$

$$g_m = \sqrt{2\mu_n C_{ox} \frac{W}{L} I_D}. \tag{5.5}$$

The corresponding large signal equations may be derived (Babanezhad and Temes, 1985)

$$I_D = k(V_{GS} - V_T)^2 \tag{5.6}$$

$$k = \frac{\mu_n C_{ox}}{2} \frac{W}{L} \tag{5.7}$$

where I_D is the drain current, V_{GS} is the gate-to-source voltage, V_T is the threshold voltage, k is the transconductance parameter, μ_n is the mobility of electrons near silicon surface, C_{ox} is the gate capacitance per unit area, W is the width of the transistor, and L is the length of the transistor.

Equation (5.6) is the simplified MOS square law characteristic in the saturation region. The equation is used to determine the large signal drain currents and the differential output current of the source coupled pair (Babanezhad and Temes, 1985)

$$I_{D1} = \frac{k}{2} \left(\sqrt{\frac{I_{SS}}{k} - \frac{V_i^2}{2}} + \frac{V_i}{\sqrt{2}} \right)^2 \tag{5.8}$$

$$I_{D2} = \frac{k}{2} \left(\sqrt{\frac{I_{SS}}{k} - \frac{V_i^2}{2}} - \frac{V_i}{\sqrt{2}} \right)^2 \tag{5.9}$$

$$I_{out} = I_{D1} - I_{D2} = k V_i \sqrt{\frac{2 I_{SS}}{k} - V_i^2} \tag{5.10}$$

where I_{D1}, I_{D2} are the drain currents of transistors M1 and M2, I_{SS} is the current of tail current source, V_i is the differential input voltage, and I_{out} is the differential output current.

Equations (5.8)–(5.10) are valid if the input voltage is

$$-\sqrt{\frac{I_{SS}}{k}} \leq V_i \leq \sqrt{\frac{I_{SS}}{k}}. \tag{5.11}$$

This is because of the non-linearity of the input pair in all differential pairs, which sets limits for the dynamic range of the input. The dynamics of the inputs may be improved by using various linearization techniques. Some linearization techniques are presented in the literature (Häkkinen, 2002; Gray and Meyer, 1993; Johns and Martin, 1997; Babanezhad and Temes, 1985; Gill et al., 1994; Gilbert, 1968). The large signal equations are useful for example when determining the output current of a Gilbert cell.

The frequency response of a source coupled pair needs to be determined in order to optimize the speed of a single transconductor to make sure that the circuit is capable of operating in the required frequencies. An additional frequency analysis needs to be performed for the circuits that are built from these differential pairs, such as for example the Gilbert cell,. This additional frequency analysis is usually made by determining the dominant pole (or zero) from the transfer function of the circuit. The dominant pole usually determines the $-3\,\mathrm{dB}$ frequency, and hence the maximum operating frequency of the circuit.

The transfer function is defined by determining the node equation of the circuit from the equivalent circuit. The determination of the poles and zeros may also be performed simply by finding the pole (or zero) frequency of each of the nodes of the circuit, in case the equivalent circuit is difficult to built (as is often the case in differential structures). This is done by using the following equation (Gray and Meyer, 1993)

$$f_n = \frac{1}{2\pi(r_n C_n)} \tag{5.12}$$

where f_n is the frequency of the node, r_n is the resistance of the node, and C_n is the capacitance of the node.

The pole, or zero, that has the lowest frequency is the dominant one (Gray and Meyer, 1993). The nodes, for example the nodes in the source coupled pair in Figure 5.15, are the nodes on the input (gates) and output (drains). This kind approximation gives a reasonable estimate of the frequency response. The actual frequency response of a circuit is found by an AC analysis.

The unity-gain frequency of an individual transistor may be determined (Johns and Martin, 1997)

$$f_T = \frac{g_m}{2\pi(C_{gd} + C_{gs})} \tag{5.13}$$

where f_T is the unity gain frequency, C_{gd} is the gate to drain capacitance, and C_{gs} is the gate to source capacitance.

As a rule of thumb, the maximum oscillating frequency of a circuitry is usually twice the f_T which determines the maximum peak frequency.

5.4.2 The Gilbert Multiplier

The Gilbert multiplier (Gilbert, 1968) is probably the most common multiplier structure used in integrated telecommunication circuits today. The basic Gilbert cell is presented in Figure 5.17. It is a double balanced (i.e. fully differential) multiplier, so it is highly immune to certain disturbances such as even order harmonics and common mode noise (Häkkinen, 2002).

In the following, the behaviour of the Gilbert cell is derived from the (large signal) drain currents of a single differential pair (Babanezhad and Temes, 1985; Gill *et al.*, 1994).

The differential output current can be expressed as (Babanezhad and Temes, 1985)

$$I_{out} = I_{D7} - I_{D8} = (I_{D3} + I_{D5}) - (I_{D4} + I_{D6})$$
$$= (I_{D3} - I_{D4}) - (I_{D6} - I_{D5}). \tag{5.14}$$

$(I_{D3} - I_{D4})$ and $(I_{D6} - I_{D5})$ are the differential currents of the upper source coupled pairs M3-M6, which form the so called mixer core (Gilbert, 1997). The differential currents can be calculated using (5.10). Note the signs of the currents in the latter term. This is due to the inverse polarity of the input voltage of the left side differential pair.

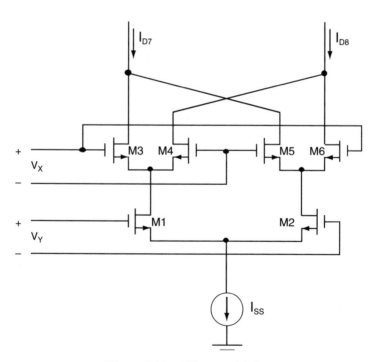

Figure 5.17 Gilbert multiplier

The tail current sources for these two differential pairs are the drain currents of the lower differential pair, which can be calculated using (5.8) and (5.9). This results in (Babanezhad and Temes, 1985)

$$I_{out} = kV_X \left[\sqrt{\left(\sqrt{\frac{I_{SS}}{k} - \frac{V_Y^2}{2}} + \frac{V_Y}{\sqrt{2}}\right)^2 - V_X^2} - \sqrt{\left(\sqrt{\frac{I_{SS}}{k} - \frac{V_Y^2}{2}} - \frac{V_Y}{\sqrt{2}}\right)^2 - V_X^2} \right] \quad (5.15)$$

where V_X, V_Y are the input voltages of upper and lower differential pairs.

It can be seen from equation (5.15) that there is a non-linear relationship between the input voltages. By limiting the input voltage swing, the output can be approximated to be linear

$$I_{out} = kV_X \left[\sqrt{\left(\sqrt{\frac{I_{SS}}{k} - \frac{V_Y^2}{2}} + \frac{V_Y}{\sqrt{2}}\right)^2} - \sqrt{\left(\sqrt{\frac{I_{SS}}{k} - \frac{V_Y^2}{2}} - \frac{V_Y}{\sqrt{2}}\right)^2} \right] \quad (5.16)$$

$$\Rightarrow I_{out} = \sqrt{2}kV_XV_Y.$$

The dynamics of the input of a Gilbert cell may also be improved by using various linearization techniques. Some linearization techniques are presented elsewhere (Häkkinen, 2002; Gray and Meyer, 1993; Johns and Martin, 1997; Babanezhad and Temes, 1985; Gill et al., 1994; Gilber, 1968).

When using Gilbert multipliers, the RF input is usually connected to the lower differential pair (M1-M2 in Figure 5.17), and the LO input is usually connected to the upper differential pair (mixer core) which acts as switches. The RF input is supposed to be differential. In case of a single-ended input, the lower differential pair may be replaced by a circuit that converts the single-ended voltage to differential current. One example of this kind of circuit is presented by Gilbert (1997). It is called a bisymmetric class AB input stage, or a micro-mixer. Figure 5.18 presents a CMOS variant of this circuit.

The single-ended RF input voltage is a.c. coupled to the diode coupled transistor (M01) on the left side. The input voltage causes an a.c. current through transistor M1 and M01. This current is inversely copied to the right side branch (M2 and M02). The operation point is set by the bias voltage provided to the gates of transistors M1 and M2.

Since the transconductance of the input stage is one term of (5.16), output current of the multiplier, the performance of the input stage is essential and has therefore been extensively studied to find better voltage to current converters. Another reason for developing new input stages is the previously mentioned non-linearity of the input pair of any differential circuit. Some of these results are briefly discussed by Häkkinen (2002).

5.4.3 Analogue Addition/Subtraction

Differential pairs can also be used to build addition and subtraction circuits. Figure 5.19 presents a schematic for analogue subtraction. The difference compared with addition

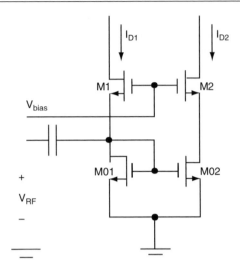

Figure 5.18   Bisymmetric class AB input stage in CMOS

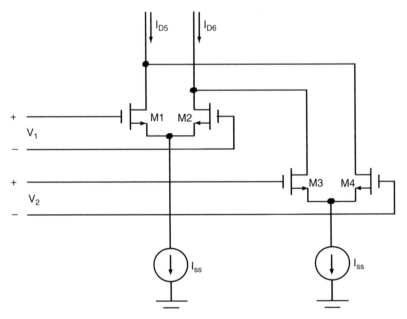

Figure 5.19   Analogue subtraction circuit

is the coupling of the drains. This kind of cross-coupling is fairly usual in differential circuits.

   Notice the similarity of the analogue subtraction circuit to the multiplier core of the Gilbert cell. However, there are two differences. First, the drains are coupled differently. This can be seen in the last term of

$$I_{out} = I_{D5} - I_{D6} = (I_{D1} + I_{D4}) - (I_{D2} + I_{D3})$$
$$= (I_{D1} - I_{D2}) - (I_{D3} - I_{D4}). \tag{5.17}$$

Second, the input voltage is different for each of the two differential pairs. These voltages are added to, or subtracted from, each other. This can be seen from

$$I_{out} = kV_1 \sqrt{\frac{2I_{SS}}{k} - V_1^2} - kV_2 \sqrt{\frac{2I_{SS}}{k} - V_2^2} \tag{5.18}$$

$$I_{out} = kV_1 \sqrt{\frac{2I_{SS}}{k}} - kV_2 \sqrt{\frac{2I_{SS}}{k}}$$
$$\Rightarrow I_{out} = (V_1 - V_2)\sqrt{2kI_{SS}}. \tag{5.19}$$

Equation (5.19) is valid in the linear region, where (Babanezhad and Temes, 1985)

$$V_1^2, V_2^2 << \frac{2I_{SS}}{k}. \tag{5.20}$$

As we can see from (5.19), the output current has a linear relationship to the difference of the input voltages. This is when (5.20) is satisfied. The use of linearization techniques may also increase the dynamic range of the input.

Equations (5.17)–(5.19) are the result of the same approach used to derive (5.14)–(5.16) for the Gilbert cell in Babanezhad and Temes (1985).

5.4.4 Integrator

The first stage of the integrator used in the implementation of the correlator is a basic structure in which a capacitor is used for the integration of current. A source coupled pair is again used as an input stage. The drains of the differential pair are coupled with a parallel capacitor as seen in Figure 5.20 (on the left side of the schematic) (Khorrama-badi and Gray, 1984). This kind of current integrator may be referred to as a fully differential Gm-C integrator, in which the source coupled pair is used as the transconductor (Gm-stage). Gm-C integrators are discussed in more detail elsewhere (Johns and Martin, 1997; Khorramabadi and Gray, 1984).

The voltage difference across the integrating capacitor is first reset by closing the switches. After a short while, when the voltage across the capacitor has reached the zero value (i.e., when both ends of the capacitor are at the same potential), the switches are opened. As the current flows through the transistors, the potentials of the drains change. If the input voltages at the transistors are different, the change of potentials in the ends of the capacitor is different in size. As a result, a voltage difference is produced across the capacitor. The voltage increases in time, and thus integrates the difference of the drain currents. The integration operation can be illustrated by (Johns and Martin, 1997)

$$V_{int} = \frac{I_{out}}{sC_{int}} = \frac{G_m V_i}{sC_{int}} \tag{5.21}$$

$$\omega_{ti} = \frac{G_m}{C_{int}} \tag{5.22}$$

where V_{int} is the output voltage of the integrator, I_{out} is the output current of the transconductor, C_{int} is the capacitance of the integrating capacitor, ω_{ti} is the unity gain frequency of the integrator, and s is the Laplace operator.

Only the voltage difference across the integrating capacitor is fed to an integrate and hold circuit, I/H (on the right side in Figure 5.20). The voltage difference is fed by an operational amplifier which acts as a switched capacitor (SC) inverting integrator (Johns and Martin, 1997). The operational amplifier must have sufficiently large bandwidth to process the output of the first integrator.

The reason for using two integrators is that the first integrator is fast enough to integrate the short pulses, but it only has a limited capacity, and therefore needs to be reset after a certain time of integration. The second inverting integrator based on an operational amplifier is too slow for the fast pulses, but fast enough to integrate the output of the first integrator after, for example, 10 pulses. The first integrating capacitor is reset again after the *sampling*. This procedure is repeated until the whole pulse train (symbol) has been integrated.

The differential voltage at the output of the inverting integrator is fed to an analogue comparator circuit which makes the bit decision. An inverter may be used after the comparator since the integration result is inverted. Both integrators are reset after providing the voltage to the comparator. The integration of the next data symbol may then begin.

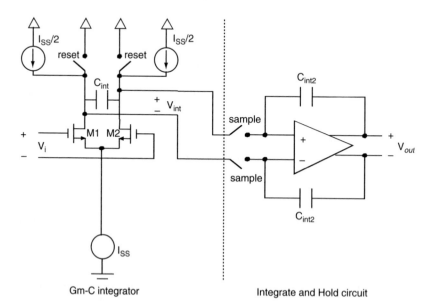

Figure 5.20   Integrator schematic

5.4.5 Current Source

Current sources are used to provide bias current for the circuits designed. A current mirror is one good way to provide bias from a global current source. The global current source has a fixed value, but the individual circuits may need more or less current. With a current mirror, the amount of current may be adjusted by changing the W/L ratio (Gray and Meyer, 1993).

Figure 5.21 presents a simple current mirror. The d.c. current fed to the transistor on the left side is 'mirrored' to the right side. The ratio of the currents on the right and left side is approximated using (5.2) and (5.3) (Gray and Meyer, 1993) as

$$\frac{I_{D1}}{I_{D2}} = \frac{(W_1/L_1)}{(W_2/L_2)}. \tag{5.23}$$

In practical implementations, the output impedance of the current source should be increased to be as large as possible. This makes it perform more like an ideal current source which does not act as a load to the circuit. Hence, the output impedance has no impact on the a.c. signals, and the a.c. signals themselves have no impact on the current level of the current source. This results in a more stable operation.

Some realizations of current sources with larger output impedance are presented in Figure 5.22. More information on these circuits can be found in the literature (Gray and Meyer, 1993; Gregorian and Temes, 1986; Johns and Martin, 1997).

5.5 IC Processes

There have been significant developments in RF IC processes during recent decades. The size of the devices and the voltage levels of power supplies have decreased, whilst the speed of the transistors has increased. The transit frequency (unity gain frequency) has increased to several tens of GHz from the ∼10 GHz silicon bipolar technology

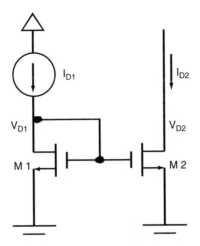

Figure 5.21 MOS current source

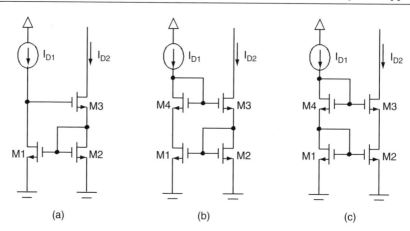

Figure 5.22 MOS current mirrors: (a) Wilson's current source, (b) improved Wilson's current source, and (c) cascade current source

prevalent in the 1980s. The upper limit of the transit frequency of a silicon homojunction bipolar transistor (for 3 V operation) is reported to be around 30 GHz for the fastest bipolar transistor today (Li, 2002).

The pure silicon bipolar process has decreased in popularity due to its low level of integration as compared with BiCMOS processes. BiCMOS combines bipolar and CMOS technologies. Digital circuits can now be integrated in CMOS and RF/analogue circuits in bipolar transistors on the same chip. In that sense, BiCMOS has the highest integration level of all IC processes, and is one of the best candidates for future RF IC processes (Li, 2002).

New technologies were developed to increase the achievable transit frequency. One of these new technologies is heterojunction bipolar technology (HBT), which is fabricated, for example, in a silicon germanium (SiGe) bipolar process. The transit frequency increases up to 50 GHz by using germanium in the base of the transistor,. Just like silicon, SiGe processes are fully compatible with mainstream CMOS process. This compatibility also allows the BiCMOS to be fabricated in SiGe. SiGe is the most commonly used RF technology at present (Li, 2002).

Another class of high transit frequency technologies is compound semiconductors (Li, 2002). The most commonly known (and used) compound semiconductors are III–V compounds (III and V are columns of the periodic table of elements), which are, for example, GaAs (gallium arsenide), GaP (gallium phosphide), InP (indium phosphide), InSb (indium antimonide) and GaN (gallium nitride). These semiconductors have physical properties such as large and direct band gap and high electron mobility. These properties make compound semiconductors suitable for high frequency applications and MMICs (microwave/millimeter-wave IC). However, compound semiconductors have a few disadvantages such as low yield and low integration level. The most commonly known GaAs processes used in RF applications are GaAS HBT and GaAS MESFET (metal semiconductor field effect transistor).

There is still much debate about which of these IC processes is the optimum choice. All of the IC processes have advantages and disadvantages. Each of the IC processes

Table 5.1   Typical figures of merit of various RF IC technologies

	Si BJT/ BiCMOS	SiGe BJT/ BiCMOS	GaAs HBT	GaAs MESFET	Si CMOS
Feature size	0.6 μm	0.25 μm	2.0 μm	0.5 μm	0.25 μm
Peak f_T	27 GHz	47 GHz	50 GHz	30 GHz	∼40 GHz
Peak f_{max}	37 GHz	65 GHz	70 GHz	60 GHz	∼50 GHz
Minimum NF@2 GHz	1.0 dB	0.5 dB	1.5 dB	0.3 dB	∼1.5 dB
$1/f$ noise corner	100–1000 Hz	100–1000 Hz	1–10 kHz	∼10 MHz	∼1 MHz
Breakdown	3.8 V	3.35 V	15 V	10 V	∼3–5 V

can be evaluated based on their properties such as transit frequency, maximum power gain frequency, minimum noise figure, $1/f$ noise corner frequency, maximum power added efficiency, linearity and reliability. Some key properties of IC processes typically used in RF applications are compared in Table 5.1 (Li, 2002).

Silicon technologies are commonly selected as they cost less and have a higher integration level. CMOS is the most available process and the dominant digital process. CMOS has higher yield and therefore lower cost per die. That gives CMOS an advantage over BiCMOS. The selective cost issues are discussed by Li (2002).

BiCMOS has advantages such as better device matching, highly accurate/predictable analogue characteristics, larger gain, and lower $1/f$ noise corner frequency. In addition, the operating life of a bipolar transistor is longer than that of a MOS (except for digital MOS circuits, which act as switches which are inactive for most of the time).

One of the operational advantages of MOS transistors is linearity, which is an important property in designing mixers or other applications. A bipolar transmitter needs emitter degeneration to achieve similar linear behaviour. Emitter degeneration requires the use of inductors or resistors, which results in decreased gain and use of larger silicon area. In UWB applications, an inductor may not be desirable, since an inductor is likely to cause a resonant frequency inside the signal bandwidth that could, for example, cause the generated pulse to oscillate.

5.6 Example Implementation

The UWB transceiver designed and discussed in this section is implemented in 0.35 micrometre CMOS. CMOS is probably the most desirable IC process, since it has the lowest cost and the best availability. CMOS is also dominant in the area of digital circuits. It is possible to avoid using multiple chips and processes by using CMOS in both RF and analogue design. As mentioned in Section 5.5, the circuits may also be implemented in other IC processes. Some of these other IC processes (e.g. BiCMOS) may also provide larger transit frequency which may result in shorter pulses.

In our example, the target pulse repetition frequency is 20 M pulses/second, and the target pulse length is less than 500 ps. This will not necessarily produce a system which is

FCC compliant. However, it will serve to demonstrate the main functional blocks and design approach for the UWB transceiver.

The transceiver implemented is part of a TH-PPM UWB system. The UWB signal is generated by a monocycle waveform generator. The generated monocycle is used for both the transmitter and receiver. Although PPM is the selected modulation scheme, BPAM is also considered in the design process.

5.6.1 Transceiver

As mentioned in Section 5.2, the architecture of the transceiver is relatively simple. The most complex part, the digital timing circuit, includes both the modulator and the pseudo-random noise coding. Together, these provide the timing for the pulse generation (see Appendix 1) and transmission. In addition, the timing circuit may be responsible for the synchronization, which increases its complexity. The synchronization is assumed ideal. The timing circuit is presented only at a schematic level and is not used in the simulations.

The transceiver components covered by this section are shown in Figures 5.24 and 5.25. They are the pulse generator (monocycle waveform generator) and the correlator at the receiver, which includes a multiplier and an integrator. Any amplifiers required in the receiver or in the transmitter are not presented. Amplifiers (or attenuators) may be needed between some circuits to provide the desired voltage swing. As an example, an attenuator is required after the pulse generator to fix the magnitude of the pulse in order to keep the transmitted power below the limits set by the FCC (2002). A low noise amplifier (LNA) is required at the front end of the receiver to amplify the received signal for the correlator, and as an operational amplifier in the integrator.

The receiver architecture has some differences in the cases of PPM and BPAM. The receiver in Figure 5.2 is for the BPAM and Figure 5.23 shows the implementation for the PPM. The PPM receiver is implemented from two correlators, which are synchronized

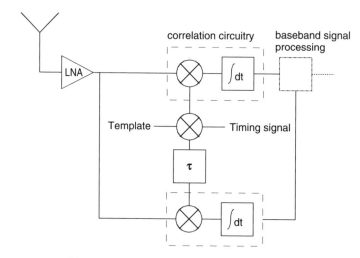

Figure 5.23   The PPM receiver architecture

to two different time windows. The first correlator detects the 'early' signal, which represents '0', and the second correlator detects the 'late' pulse, which represents '1'. The timing for the correlators is provided by a timing circuit which is similar to the timing circuit in the transmitter. The correlator is synchronized to the received signal. The timing signal is delayed by the time of the PPM modulation index (δ) for the second correlator, which is therefore synchronized to the late pulse. The bit decision is made by determining which of the correlators produces the largest output.

The BPAM receiver has only one correlator. This correlator is synchronized to the only time window set by the PR code. The bit decision is made from the output of the correlator. The output of the correlator depends on the polarity of the transmitted pulse.

5.6.2 Pulse Generator

One method of producing a monocycle is to sum up two pulses with different polarities with a time delay corresponding to the length of the pulses. This may be done in a number of different ways. One method was introduced in an earlier Section 5.4 where a transmission line technique was utilised (Lee *et al.*, 2001a). Another technique was introduced in Section 5.3.4 in which two pulses with a time delay were combined in a multiplier so that one of the pulses was multiplied by 1 and the other pulse was multiplied by -1 (McCorkle, 2001). By changing the multiplier's control signals, the phase of the produced monocycle may be changed 180 degrees, and the multiplier acts as a binary pulse amplitude modulator.

In the case of PPM, the modulator structure in McCorkle's work (2001) is not required, so the circuit may be simplified. Such a circuit is presented here (Tiuraniemi, 2002). The presented pulse generator subtracts two differential pulses with a time delay corresponding to the length of the pulses from each other, thus producing a differential monocycle. This is done by an analogue subtraction circuit which is much simpler than any multiplier. BPAM may also be utilized even though the advantages of this pulse generator mainly concern a PPM system.

The top level schematic of the pulse generator is presented in Figure 5.24. The numbers in brackets indicate the waveforms in different phases of the pulse generation shown in Figure 5.25.

Two short pulses are generated with a digital pulse generation circuit presented in Figure 5.26 and Figure 5.27. The delay between the two pulses is realised by a simple delay element consisting of two inverters and a PMOS varactor, which provides the opportunity to tune the length of the delay. These short pulses are then fed as inputs to a single-ended to differential converter presented in Figure 5.28. The resulting differential pulses are fed to the inputs of two emitter or source coupled pairs of the analogue subtraction circuit (see Figure 5.29). The outputs of these differential pairs are cross-coupled so that a monocycle is produced.

Figures 5.26 and 5.27 present two methods of generating short pulses. The first method utilizes an XOR gate and the second method utilizes an AND gate. The XOR operates so that whenever both input signals are at different *logical levels*, the output is at a high level, i.e. *logical '1'*, and when both input signals have the same *logical levels* the output is at low level, i.e., *logical '0'* (Daniels, 1996). The length of the high-level output can be adjusted by the phase difference between the inputs. This is also provided by a simple delay element.

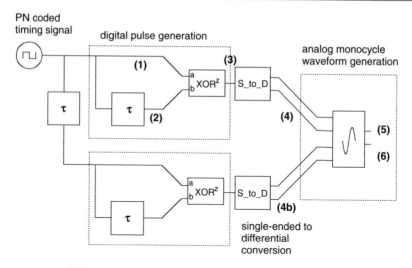

Figure 5.24   Top level schematic of pulse generator

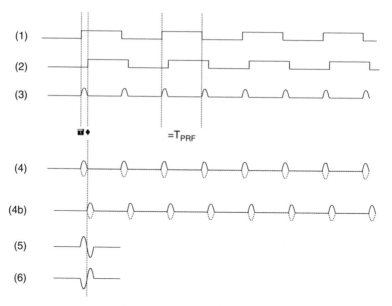

Figure 5.25   Waveforms in different phases of monocycle generation presented in Figure 5.24

The AND-gate operates so that whenever both the inputs are at a high level, the output is at a high level (Daniels, 1996). An inverter is connected to one of the inputs of the AND gate, and a clock signal is connected to both the inverter and the AND gate. The output of the AND gate is now at low level at all times, but the output still reacts to the rising edge of the clock. This is because the inverter has a delay during which its output is still at

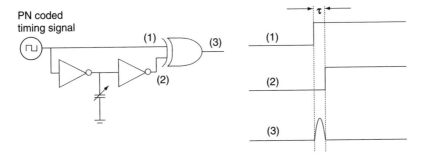

Figure 5.26   Digital pulse generation by an XOR-gate

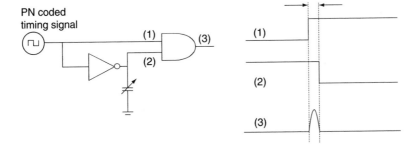

Figure 5.27   Digital pulse generation by an AND-gate

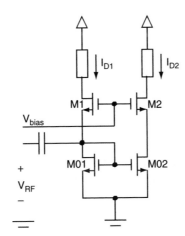

Figure 5.28   Single-ended to differential converter

zero while the clock has already risen. This effect is called a glitch, or a hazard, which results in a short pulse with length corresponding to the length of the inverter's delay. The delay can be adjusted by connecting a varactor from the output of the inverter to ground. If the delay is very short, the resulting pulse might not have time to rise all the way up to the *logical "1"*, but it is still useful for generating a short pulse. When using this circuit, the

PRF decreases to half that when using the XOR-circuit, since the short pulse will occur only on the rising edge of the clock. The XOR reacts to both the rising and falling edge of the clock. This effect can also be seen from Figure 5.25.

A NAND-gate may also be used. The operation is the same as in case of an AND gate, the difference being that the pulse will occur on the falling edge of the clock and the pulse is from high to low level. The NAND-gate is faster than the AND-gate and is therefore used in the implementation of the transceiver. The shape of the pulse is defined by the RC time constant of the used gate in all of the three circuits.

Two pulses separated by a controllable time delay are produced by combining two of these digital pulse generators, with the input to one being delayed. These two pulses are fed to the single-ended to differential converter to produce two differential pulses. One example of such a converter is a transconductance stage, which inversely copies the AC current (caused by the AC-coupled input voltage) from the left branch to the right branch (see Figure 5.28). The two currents are then turned into voltage swings by the resistors. This circuit is a variant of a BJT micro-mixer (Gilbert, 1997) and was introduced in Section 5.4.2.

The two short differential pulses that have been generated are now fed to the differential pairs, which are presented in Figure 5.29. The outputs of the differential pairs are cross coupled so that a differential monocycle (see Figure 5.25 for the theoretical waveforms) is produced as a result of linear subtraction operation.

The differential pairs may consist of bipolar transistors or FETs. The main difference between these two types of transistor is the operating speed, which is supposedly faster for a bipolar transistor. Other transistors such as HEMT and HBT can also be used.

The load impedance of the differential pairs consists of resistors or an active PMOS load. The active load increases the complexity and the number of transistors, but decreases the used chip area. Active loads are commonly used in CMOS designs. One

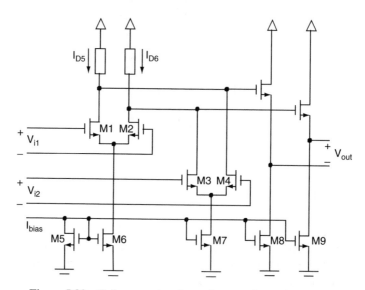

Figure 5.29 Pulse generator's analogue subtraction circuit

advantage of active loads is that a very large impedance may be realized using a small area of silicon. An active load is not necessarily suitable in a high frequency application, since it is known to have poor high frequency characteristics. Therefore, resistive loads are used in this example. Using an inductor may not be feasible since it may lead to the existence of a resonant frequency inside the signal bandwidth, which may make the circuit oscillate.

5.6.3   The Analogue Correlator

The multiplier of the analogue correlator used in this implementation is the Gilbert multiplier cell. The choice is justified by the benefits explained in Section 5.4.2. Another reason is the possible implementation of the BPAM in which only the polarity of the received signal needs to be detected. Since a Gilbert cell is a four-quadrant multiplier, i.e. the result of the multiplying may be -1 (opposite polarization), $+1$ (equal polarization) or zero (no correlation), the Gilbert cell may be used to detect different polarities.

The integrator is realized by a simple differential pair with a capacitor connected parallel to the drains. The voltage across the capacitor increases as the result of the multiplication is fed to the integrator. Every time a pulse is received, the voltage across the capacitor increases in proportion to the input of the differential pair. The polarity of the voltage depends on the polarity of the product of the multiplication. The polarity of the product of the multiplication depends on the polarity of the received pulse. This correlator, presented in Figure 5.30, is therefore compatible in both cases of BPAM and PPM. In PPM, two of these correlators are used so that the first correlator detects the early pulses and the other correlator detects late pulses.

After the Gm-C integrator, an integrate and hold (I/H) circuit is used to improve the performance of the integrator. The I/H circuit is realized by a discrete time inverting integrator built from an operational amplifier.

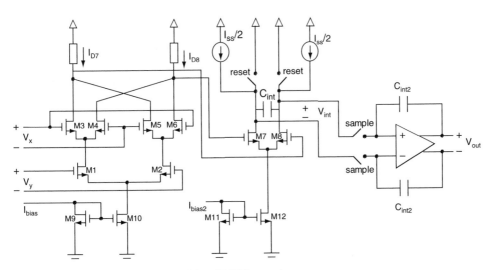

Figure 5.30   CMOS correlator structure

The monocycle of the transmitter is used as the template waveform in the multiplier. Even though the received waveform has been distorted from the transmitted monocycle, the correlation still produces useful results. The correlation result is not as good as it would be in the case of autocorrelation (in which the signal is multiplied by itself). The mismatch decreases the SNR of the correlating receiver however the performance is still adequate.

5.6.4   Timing Circuit

A timing circuit provides the trigger signal (or timing signal) for the pulse generator. The timing circuit is responsible for the PN coding and modulation in the case of PPM. The top level schematic of the timing circuit is presented in Figure 5.31.

The PN code is provided by a pseudo-random sequence generator, which may consist, for example, of D flip-flops and an XNOR-gate (Daniels, 1996). This is presented in Figure 5.32. The operation is such that the generator is given an initial state, which may be any state except all '1's. The generator steps in pseudo-random order from the initial

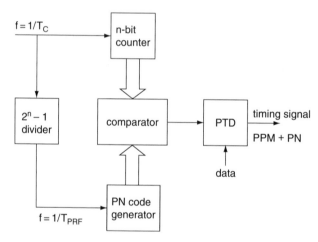

Figure 5.31   Top level schematic of timing circuit for generating TH-PPM signals

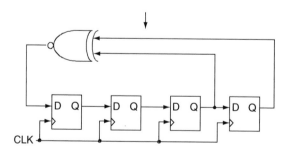

Figure 5.32   Pseudo-random sequence generator

Table 5.2 Results of the processing gain calculations for the implementation

T_{PRF}	T_C	ΔT	N_U	N_S	PG
50 ns	1.6 ns	0.8 ns	31 (5 bits)	200	41 dB

state through all the states, excluding the 'all 1' state, as these states cannot be generated by the loop. In addition, the loop cannot continue from the 'all 1' state due to the operation of the XNOR-gate. Therefore the loop may not be initiated in that state.

Each time frame set by the PRF generates a new code. The number of states is determined from the number of memory elements (i.e. the number of bits, n) to give a sequence of length $2^n - 1$ bits. The number of states corresponds to the maximum number of unaliased users operating simultaneously in the network.

Any synchronous counter may be used as the counter in 5.31. The counter steps through all the states starting from zero. When the counter reaches the last state, it recommences from the beginning. The time spent for one cycle (count through all the states) equals the length of the time frame. The number of states also matches the maximum number of users.

The digital comparator in Figure 5.31 compares the PR code to each of the states of the counter. A TRUE (i.e. *logical '1'*) signal is produced whenever the PR code and the states of the counter match. The TRUE signal occurs once in every time frame set by the PRF. These pulses form the PR coded timing signal.

In case of PPM, this timing signal is fed to a programmable time delay (PTD). The PTD either delays the signal by ΔT in case of '1', or lets it through non-delayed in case of '0'. This results in a TH-PPM signal which is provided to the pulse generation circuit as the timing signal. There is no need for the PTD in the case of BPAM.

The same kind of timing circuit, without the PTD, may also be used in the receiver to provide timing for the correlator. However, the delay experienced in the medium between the antennas must be taken into account.

The processing gain calculations given in Table 5.2 provide results that are adequate for this implementation. The results are presented in Table 5.2. The length of the PR code, time frames, slots and delays (in seconds) and the pulse repetition frequency are achieved from these results.

5.7 Simulation Results

In this section, the simulation results are presented and discussed. First, the functionality of the pulse generator and correlation receiver is confirmed by a transient analysis. Second, a corner analysis is run for both the transmitter and receiver to evaluate the performance of the system with respect to variations in process, temperature and power supply. A commercially available simulator, Spectre, is used for the simulations. Spectre is a Cadence version of the SPICE circuit simulator. The top level schematics of the simulated circuits are depicted in Appendices 3–10.

The effects of each of the antennas and the channel are not taken into account. This means that the transmitted signal is received without distortion or delay. The performance of the correlator is more readily evaluated in such an arrangement. The correlation between the template waveform (transmitted waveform) and the received waveform was presented in Figure 5.3.

5.7.1 Transmitter

At this point, the timing circuit was not implemented and the timing for the pulse generation was realized with a clock signal. The simulations of modulation and PN coding are left for future work. The results of transient and corner analysis of the transmitter (Appendix 3) are presented below.

The length and shape of the generated monocycle primarily depends on the digital pulse generation (Appendix 4) circuit which produces the short pulse utilized in the monocycle generation. The length of this short pulse is approximately one half of the length of the monocycle, and may be adjusted by a PMOS varactor. However, the varactor should be left out to minimize the delay, so that the shortest possible pulse width is achieved. The shape of the short pulses is nearly Gaussian (or half of a sine period) due to the nature of the glitch. Mathematically, the generated monocycle is not equal to the first derivative of the Gaussian pulse. However the waveform may be considered as a reasonable approximation to the first derivative.

The minimum pulse width is set by the delay of the inverter, assuming the NAND-gate does not introduce additional delay. The use of a NAND-gate in the circuit leads to faster operation compared with the use of AND- or XOR-gates. As seen in Figure 5.33, the delay between the clock signal and the inverted clock signal is about 250 ps, and the pulse width is about 280 ps. From this it can be seen that the NAND-gate has a small effect on the pulse width. The pulse width of 280 ps is near the minimum pulse width with this kind of a circuit in this process (CMOS 0.35 μm). Shorter delays must be generated for shorter pulses. This may be achieved by having more inverters of different speed connected parallel to take advantage of the differences of the delays. Another approach is to have faster NAND-gates and inverters available in another process.

Figure 5.34 shows the result of single-ended to differential conversion. The pulse width is not affected by the circuit. The differential pulses generated with the single-ended to differential conversion circuit (Appendix 5) can be connected directly to the input of the pulse generator. However, an attenuator can be useful to adjust the amplitudes of the input signals, since the peak-to-peak voltage of the differential pulses exceeds the dynamic range of the input pair of the pulse generator. By attenuating the pulses, the pulse generator operates in the linear range.

The monocycle generation is presented in Figures 5.35 and 5.36. The first figure shows the input and output signals as single-ended signals to give a clear idea how the monocycle waveform generation circuit (Appendix 6) works. The input signals are marked as negative and positive pulses. By cross-coupling these pulses, two monocycles with 180-degree phase difference are generated. The monocycles are not identical due to some differences in the amplitudes of the negative and positive pulses of the output. However, this is not a problem in a differential structure, since it is the difference of the single-ended outputs that matters.

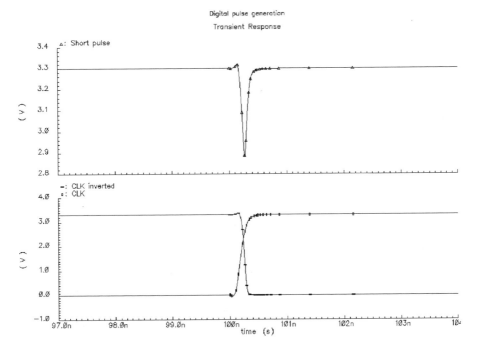

Figure 5.33   Digital pulse generation

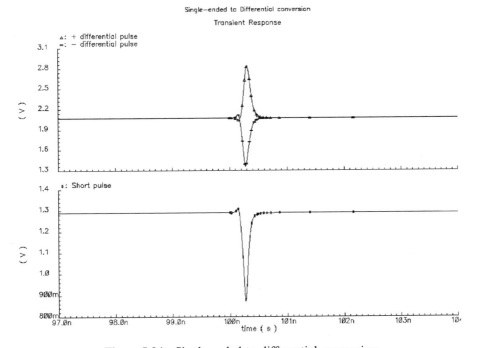

Figure 5.34   Single-ended to differential conversion

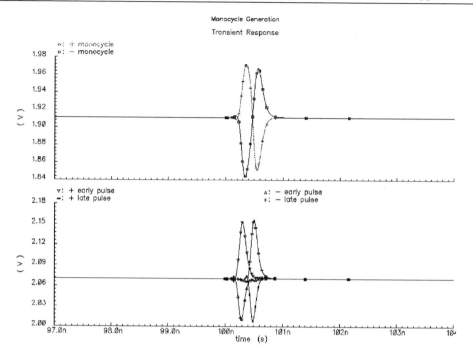

Figure 5.35   Monocycle generation (single-ended signals)

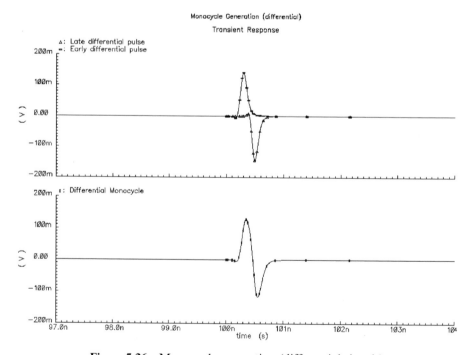

Figure 5.36   Monocycle generation (differential signals)

The differential inputs and output are depicted in Figure 5.36. Here the cross coupling may be presented as a difference between two differential pulses, as seen in the upper curves. As a result, a monocycle is generated. The pulse width of the monocycle is approximately 650 ps. This pulse width indicates that some additional delay is introduced by the analogue pulse generator. By optimizing the circuit design, the pulse generator may have a higher operating frequency and therefore would not introduce additional delay. The delay between the two branches providing the input for the waveform generator also affects the final pulse width. The pulse width may be further shortened by minimizing this delay.

The pulse width corresponds to a centre frequency of about 1.54 GHz. The transmitted signal is not the same since it is differentiated due to the effect of the antenna.

The method of producing the monocycles is not perfect. A pulse is intended to be produced only on rising (or falling) edge. In Figure 5.37, which presents a pulse train, small imperfections can be seen between the monocycles. These are the undesirable products of the digital pulse generation that occur when the clock signals change states. By increasing the delay, the short pulse (glitch) increases in both amplitude and width, and the undesirable product decreases. In other words, the problem is solved by lowering the operational frequency. The problem is also solved by using the XOR-gate in the digital pulse generation, since every change in the clock signals is utilized. The XOR-gate would also increase the pulse width due to its slow operation. Such problems are common in digital circuits that are pushed to their limits. This behaviour could also be avoided by having faster digital cells in the process. Another way to remove these undesirable effects is to use some kind of windowing technique to cut them off.

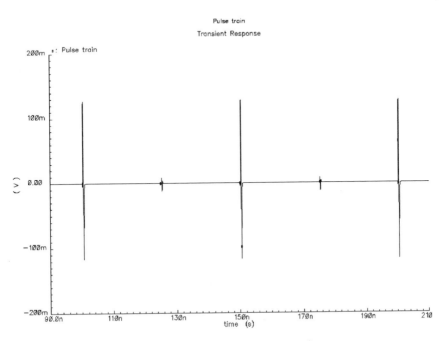

Figure 5.37 Monocycle pulse train

The waveform after the transmitting antenna is depicted in Figure 5.38. It is a derivative of the generated monocycle, since the antenna acts as a differentiator. The differentiation is performed by the calculation tool in Cadence. The waveform may be compared to the third derivative of the Gaussian pulse.

The power spectral density (PSD) of the transmitted signal (i.e., after the transmitting antenna) is presented in Figure 5.39. The PSD was also produced using the calculation tool in Cadence. The amplitude is normalized to 0 dBm. More important information is the bandwidth and centre frequency. The centre frequency is 2.4 GHz, and the −3 dB bandwidth is 1.8 GHz. This does not satisfy the FCC frequency mask (Fig 1.1). The pulse width may be shortened and the FCC mask satisfied by using a faster IC process to obtain reduced delays and higher operational frequency. The spectrum of the signal may also be modified by filtering. This would lead to the desired spectrum, but the waveform would be distorted.

In the high frequencies, from 5 to 8 GHz, a second maximum is seen. This is supposedly caused by the undesirable products of the digital circuitry. The power level of this spurious signal is more than 10 dB below that of the desired signal, so it may be considered a non-harmful component. In any case, the second maximum is placed inside the FCC's frequency mask (−41.25 dBm/MHz at 3.1–10.6 GHz). Even if the spectrum is shifted 2 GHz to the right on the frequency axis (which would make the spectrum FCC compliant), the power level of the unwanted side band is low enough to satisfy requirements.

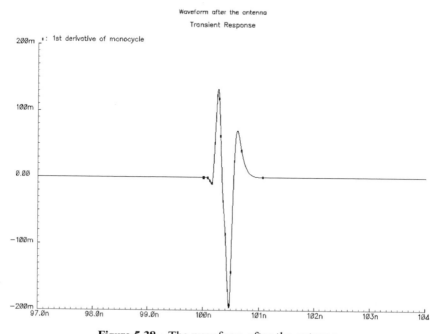

Figure 5.38 The waveform after the antenna

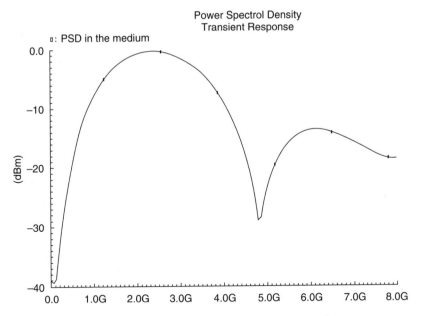

Figure 5.39 Power spectral density of the pulse train after antenna

From the presented results, some characteristic figures may be calculated using equations presented in Chapter 3 symbol time and rate are

$$T_S = N_S T_f = 200 \cdot 50\,\text{ns} = 10\,\mu s$$

$$R_S = \frac{1}{T_S} = 100\,\text{kbits/s.}$$

Symbol rate could easily be increased by transmitting a smaller number of pulses per symbol or having a smaller duty cycle, which may be achieved by increasing the T_f. The trade-off is a decreased processing gain, PG, which is now

$$PG = 10\log N_S + 10\log \frac{T_f}{T_p} = 23\,\text{dB} + 19\,\text{dB} = 42\,\text{dB}$$

The fractional bandwidth of the transmitted signal is

$$B_f = 2\frac{f_H - f_L}{f_H + f_L} = 2\frac{3.3 - 1.5}{3.3 + 1.5} = 0.75 > 0.20$$

from which it is seen that it may be said to be an UWB signal. The $-3\,\text{dB}$ bandwidth of the UWB signal is $1.8\,\text{GHz}$. The bandwidth expansion factor is as high as

$$B_e = \frac{B}{R} = \frac{1.8\,\text{GHz}}{100\,\text{kbit/s}} = 18\,000 = 43\,\text{dB,}$$

which is a large value. However, it is clearly expected bearing in mind that the data rate of the designed system is sufficiently low compared with the possibilities of a UWB system. A 2Mbps UMTS system has an expansion factor of $2.5 = 4\,\mathrm{dB}$. The UWB system of this chapter would have an expansion factor of $900 = 30\,\mathrm{dB}$ for a bit rate of 2 Mbps.

5.7.2 Receiver

The receiver (Appendix 7) was also simulated using the above-mentioned analysis. The timing and synchronization for the correlator was provided by the transmitted pulse train, which was used as both the received signal and template waveform (autocorrelation). The results of transient and corner analysis are presented in the following sections.

Figures 5.40 and 5.41 depict the inputs and outputs of the Gilbert multiplier (Appendix 8). The polarity of the received signal (RF input) is different in the two figures. As may be seen, the output of the multiplier depends on the polarity of the received signal. When the polarity of the received signal is the same as the polarity of the template waveform, the output is positive, and when the polarity of the received signal is opposite to the polarity of the template waveform, the output is negative. This also applies to the output of the correlator.

The first integrator integrates every received pulse. As seen in Figure 5.42, the voltage across the integrating capacitor increases by a certain amount every 50 nanoseconds,

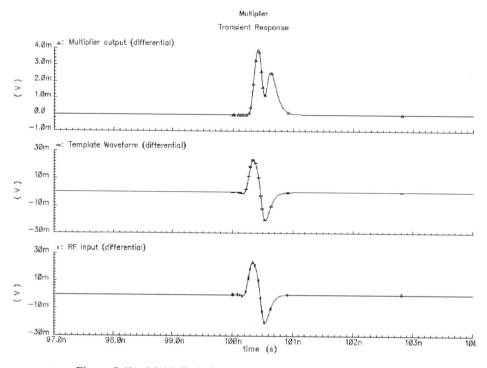

Figure 5.40   Multiplier's inputs and output (differential signals)

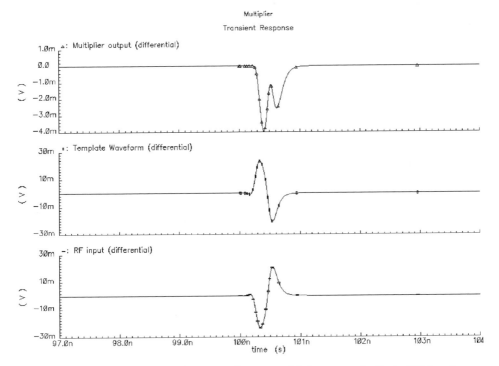

Figure 5.41 Multiplier's inputs and output with inverse polarity in RF input

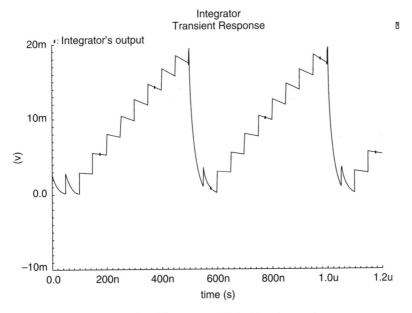

Figure 5.42 The output of the first integrator

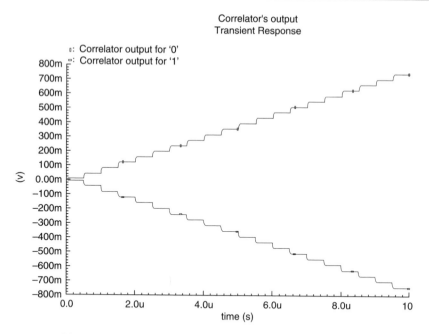

Figure 5.43   The outputs of the correlator in case BPAM

which is the pulse interval. After integrating a few pulses, the efficiency of the first integrator starts to decrease. The voltage across the capacitor increases in a normal way, but it is no longer able to hold its level. To avoid the loss of integrated energy, the voltage is sampled and then reset. This procedure is repeated until the whole pulse train is integrated. This takes 10 microseconds (200 pulses times 50 ns) with the used simulation parameters.

The sample is further integrated in the inverting integrator, which also provides some extra gain. The output of the second integrator is depicted in Figure 5.43. The output increases after each period of 500 ns (10 pulses). After the symbol is integrated, the output of the correlator has reached a value that is detectable by semiconductor devices, in this case a comparator.

The curves in 5.43 represent the outputs in case of '0' and '1' in BPAM. The output for PPM is the same as '0' for BPAM.

5.8   Conclusions

This chapter has explored some of the fundamental issues that need to be addressed when attempting to implement an UWB transceiver. Various circuit fundamentals were examined as were commercial solutions and solutions described in the open literature.

The chapter focuses on the main aspects of the UWB transceiver; the pulse generation and key aspects of the receiver architectures. Several pulse generation techniques were explored, including a digital generation technique.

The chapter highlighted some of the properties of the CMOS process which must be taken into account when an UWB design is fabricated on a wafer. Parasitic components and other non-ideal characteristics are not modelled with adequate accuracy in typical commercial CMOS design tools. The parasitic components must be modelled by the designer, after which the transceiver must be re-simulated. The parasitic components are likely to have a significant impact at the frequencies of interest. The parasitic components must be avoided by careful design selection. The parasitic components and non-ideal characteristics are usually modelled more accurately in bipolar technologies. Some cases still need to be modelled to verify the practical behaviour. The best way to determine the behaviour is to measure the fabricated IC.

An example UWB transceiver implementation was shown based on relatively straightforward CMOS technology. The pulse generation circuit developed was verified as a very efficient and moderately reliable way to generate a monocycle. The example UWB transceiver implementation provided an insight into the design issues for an UWB transmitter, even though the transmitted signal did not satisfy the limits set by the frequency mask specified by the FCC. There are at least two possible ways to try to change the spectrum of the transmitted signal in the circuit example. The first way is to have a shorter pulse, which may be realized by using a different pulse generation technique, or different IC process. The second way is to shape the generated pulse using a pulse-shaping filter so that the mask is satisfied. The latter method is much more promising in practice, however is more likely to reduce the efficiency of the receiver.

One of the practical issues to consider when implementing UWB circuits is the realization of synchronization. The realization of synchronization may be difficult to achieve due to the ultra-short pulses and their fast rising and falling times. In this chapter, the synchronisation was assumed ideal.

The integrator structure developed in the chapter demonstrated excellent performance in integrating fast pulse with fast rising times. The first stage of the integrator is very sensitive and is able to integrate short pulses with more than adequate results. The second stage provides additional capacity for the integrator to make long integration times possible in cases where the received signal has low energy.

6

UWB Antennas

Tommi Matila, Marja Kosamo, Tero Patana, Pekka Jakkula,
Taavi Hirvonen, Ian Oppermann

6.1 Introduction

Whilst often ignored or assumed ideal in conventional narrower band system analysis, antennas are a critical element in the signal flow of UWB systems. The antenna acts as a filter for the generated UWB signal, and only allows those signal components that radiate to be passed. The antenna is often approximated as a differentiator both at the transmitter and receiver. This is the motivation behind the examination of the various derivatives of the Gaussian pulse explored in Chapter 3.

The goal of this chapter is to explore antenna structures suitable for use in short range, low power indoor UWB radio systems, and outdoor 'base station' communications. This chapter examines some of the main UWB antenna types (dipoles, bow-tie and TEM horn), as well as discussing the performance of a small class of antennas that are suitable for low-cost communications systems. Various practical antenna structures are examined through simulation and measurements results of several prototypes are presented. It is outside the scope of this work, however, to undertake a rigorous theoretical performance evaluation of the antenna types.

6.2 UWB Antenna characteristics

The UWB antenna acts like a filter and is a critical component in UWB radio systems. The basic effect of antennas is that they produce the derivative of the transmitted or received pulse waveform (Funk *et al.*, 1995a). This also has the effect of extending the duration of the transmitted and received pulse. This extension of pulse duration decreases the time resolution of the system. The antenna has a greater impact in UWB than in narrower band systems because of the very large bandwidth of an UWB signal.

UWB Theory and Applications   Edited by I. Oppermann, M. Hämäläinen and J. Iinatti
© 2004 John Wiley & Sons, Ltd   ISBN: 0-470-86917-8

In antenna terminology, the frequency range demand must be 6:1 or greater in order to be 'ultra wide' (Taylor, 1995) which means that the upper frequency must be at least six times greater than the lower frequency of the band. For such very wideband antennas, issues of linearity, radiation efficiency and impedance match across the band present difficult problems.

One problem which will arise when a very short time domain (im)pulse (implying large bandwidth) is used to excite the antenna, is the ringing effect. After the antenna, the signal is no longer impulse like. Instead the pulse is spread in the time domain. A typical antenna response is presented in Figure 6.1, where the ringing effect is modelled using a simple Bessel function.

To avoid ringing, resistive antennas with low Q-values should be used. The resistive loading will cause the unwanted signal component to die away quickly, leaving a pulse much closer to the desired shape. The antenna bandwidth can also be increased by making the Q-value small, since the bandwidth is inversely proportional to Q-value. However, the low Q-value implies that the efficiency of a resistive antenna is generally quite poor.

The Q-value for an antenna is given by

$$Q = f_o/(f_H - f_L)$$

where f_o, f_H, and f_L are the centre frequency and the upper and lower 3 dB frequency values of the antenna respectively.

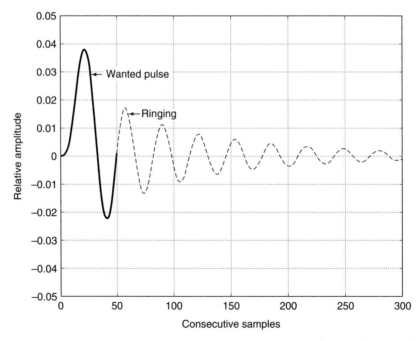

Figure 6.1   The response of an antenna to impulse excitation showing the ringing effect

The frequency domain is also useful for describing the transient response of antennas because the time and frequency domain are connected by the Fourier transform. The ability of an antenna to preserve the waveform of the ultra-narrow pulse is investigated in the time domain. Two of the most important time-domain properties of an antenna are fidelity and symmetry. The fidelity is defined as the maximum cross-correlation of the normalized incident voltage and the normalized electric field in the far field region (Montoya and Smith, 1996). The symmetry is a measure of the symmetry of the waveform in the far field region (Montoya and Smith, 1996). More theoretical approaches on time-domain antenna characterization can be found in the literature (Balanis, 1997; Montoya and Smith, 1996; Shlivinsky *et al.*, 1993; Allen *et al.*, 1993; Lamensdorf and Susman, 1994).

UWB antennas differ from their narrowband counterparts in one basic concept. Many antennas, especially in the telecommunication applications, are resonant elements that are tuned to particular centre frequencies and have relatively narrow bandwidths. In contrast, UWB antenna designs seek much broader bandwidths and require non-resonating operation.

In the literature, several antenna types have been presented for use in UWB systems. Applications have used dipoles, log-periodic dipole arrays (LPDA), conical monopole, spirals, notched, ridged and TEM horns antennas (Taylor, 1995). The main focus in antenna technology in the existing UWB literature, is for high power radar antennas. Table 6.1 presents a summary of typical dimensions of different antenna elements (Taylor, 1995). The remainder of this chapter will explore the practical implementation aspects of several antenna types.

6.3 Antenna types

In this section, the antenna types suitable for UWB operation are described. In Section 6.3.1 the general requirements for antennas are described. Sections 6.3.2 to 6.3.8 concentrate on the different antenna types.

6.3.1 General Requirements

This section evaluates the transient response of antennas for base-station and indoor-portable use. The short, sub-nanosecond pulses require special antenna structures compared with typical narrowband systems. An ideal wideband antenna acts like a high-pass filter, which means that the pulse waveform is differentiated when passing through the antenna. Typical wideband antennas, such as log-periodic dipole arrays, change the waveform even more due to dispersion. In general, UWB antennas should be linear in phase and should have a fixed phase centre. Typical impedance circuits may not be phase linear. For this reason the antennas should be inherently impedance matched. The antenna's radiation characteristics also have a significant impact on the antenna's performance. The antenna gain should be smooth across the frequency band in order to avoid dispersion of the transmitted pulse. The antenna gain will typically appear different from different angles. This will lead to different pulse shapes depending on the angle to the receiver.

Table 6.1 Target antenna values

	Base station antenna	Portable antenna
Frequency	2–10 GHz	3–10 GHz
Matching	typical VSWR < 1.5	typical VSWR < 2
	VSWR < 2	VSWR < 3
Radiation efficiency (min.)	50%	10%
Directivity (typical)	0–30 dB	0 dB
Fidelity (typical)	> 0.7	> 0.7
Size	not specified	area < 100 cm^2 on PCB

VSWR is voltage standing wave ratio.

Some target antenna specifications for base station and portable antennas are given in Table 6.1. These are by no means definitive. However they give an indication of appropriate parameter values.

6.3.1.1 Base Station Antenna

As discussed in previous chapters, the power of UWB transmitted signals is extremely low. In this context, 'base station' antennas may be used for networks such as high speed data kiosks or for low data-rate systems, including location and tracking systems. The base station antenna may be designed for indoor or outdoor usage, depending on the application. Outdoor usage allows the antenna to be relatively large, with dimensions of several wavelengths. In addition to the requirements outlined in Table 6.1, the antennas must radiate efficiently. This may restrict the use of resistively loaded structures. Base station antennas may be either directive or omnidirectional. Directional antennas could be used for example in radio links, whereas omnidirectional antennas would be more favourable in mobile applications.

6.3.1.2 Portable Antenna

The requirements for the portable, short-range, UWB antenna differ slightly from the base station antenna. Most importantly, the antenna must be small. It is also highly desirable that the antenna be low cost and preferably constructed on a printed circuit board. The small size implies that the antenna is omnidirectional. The radiation efficiency is not as critical parameter as in base station antennas, which makes it possible to use resistive loading. If possible, the transceiver will be embedded in the same circuit board as the antenna.

6.3.2 TEM Horn

The TEM horn and its variations are among the most commonly used antennas in UWB applications. The basic structure consists of two tapered metal plates fed by a two-wire TEM-mode transmission line. The structure can be considered as a bent bow-tie dipole. The TEM horn preserves the pulse waveform very well and has a constant phase centre (van Cappellen *et al.*, 2000). The flaring and length of the antenna can be adjusted to modify the radiation pattern, impedance matching, and the transient behaviour of the antenna (Shlager *et al.*, 1996). The gain of the TEM horn ranges from 5 to 15 dB, which is suitable for directive base-station operation.

The spatial dispersion caused by the non-planarity of the waveform from TEM horn can be corrected by the use of lenses (Baum and Stone, 1993) or reflectors (Foster, 1993; Baum and Farr, 1993). One such antenna is the impulse radiating antenna (IRA) (Baum and Farr, 1993) which will be described in Section 6.3.4.

6.3.3   TEM Horn Variants

There are several modifications of the TEM horn. Septum plates can be put inside the horn to improve the field uniformity (Buchenauer *et al.*, 1999). Another technique described by (Yarovoy *et al.*, 2000) is to fill the horn with dielectric material. One half of the horn can also be replaced by a large ground plane such that the horn resembles a bent bow-tie monopole. In this last case, a broadband balun is not required. It is also possible to use resistive loading in the antenna to suppress the reflections from the end of the horn (Shlager *et al.*, 1996).

The idea behind a ridged horn and a tapered slot antenna (Yngvesson *et al.*, 1989; Lewis *et al.*, 1974) is similar to that of the TEM horn: a TEM- or quasi-TEM mode is used to feed the antenna. With a ridged horn structure, there is a ridge inside the TEM horn. This approach appears to yield more constant gain with respect to frequency relative to standard TEM horns. However, the structure is more difficult to produce. In addition, the antenna appears not to be optimal for pulse radiation (van Capellen, 1998). The polarization diverse antenna by Wicks and Antonik (1993) can also be considered as a variation of a ridged horn.

Tapered-slot antennas can be constructed using printed-circuit-board techniques. With such an antenna, however, it may be difficult to achieve constant gain. The tapered-slot antenna and its variations appear to be good candidates for portable antennas. Another method for constructing TEM-horn-type antennas on printed circuit boards is described by Nguyen *et al.*, (2001).

6.3.4   Impulse Radiating Antenna

An impulse radiating antenna (IRA) consists of a TEM horn feeding a parabolic reflector (Baum and Farr, 1993). With such an antenna, it is possible to obtain high and almost frequency-independent gain. The reported gains are of the order of 25 dBi. The antenna gain can be made adjustable (Farr *et al.*, 1999) by moving the TEM horn feed off the focal point of the parabolic reflector.

The high gain makes this antenna a good candidate for very long range base-station applications. The high gain, narrow beamwidth and short pulses generated by this antenna make it highly immune to interference.

6.3.5   Folded-horn antenna

A folded horn antenna for UWB high power applications is presented by Kardo-Sysoev *et al.*, (1999). The idea of the folded-horn antenna comes from sub-horns inserted in

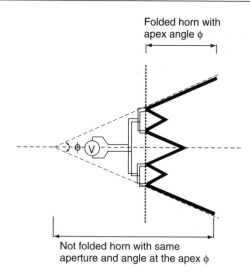

Figure 6.2   Folded horn antenna

a main horn. The sub-horns divide the initial horn aperture into two equal parts. Using this technique, the size of the antenna can be reduced. This can be seen from Figure 6.2. which shows the structure of folded horn antenna.

6.3.6   Dipoles and Monopoles

Dipoles and monopoles without resistive loading are based on resonance techniques, therefore the ringing effect described earlier significantly degrades their performance. A comprehensive review of differently loaded thin-wire monopoles is presented by Montoya and Smith (1996). The results indicate that none of the monopoles examined preserve the pulse characteristic. The resistive loading, however, seems to improve performance significantly.

A popular design is the bow-tie antenna (also called the 'bifin' antenna, Stutzman and Thiele, 1981). The bow-tie antenna is used in an ultra-wide antenna design Lai *et al.*, (1992) which shows some examples of different ultra-wideband antennas.

The beamwidth and input impedance of a bow-tie antenna depend directly on the antenna geometry, and they are nearly constant over the desired frequency range. To ensure a balanced and wideband feed to the bow-tie antenna, a hybrid construction with a slot line antenna is used (Lai *et al.*, 1992). The bandwidth of a bow-tie antenna depends on the length of the plate (see Figure 6.3). The flare angle and the length of the plate define the lower frequency. The beamwidth of the bow-tie antenna varies linearly with the flare angle.

The efficient use of resistive materials also appears to be beneficial in the bow-tie antennas (Maloney and Smith, 1993a; Shlager *et al.*, 1994).

There are other designs that can be considered as dipole structures, such as the resistively loaded antenna by Chevalier *et al.* (1999) optimized for pulse radiation.

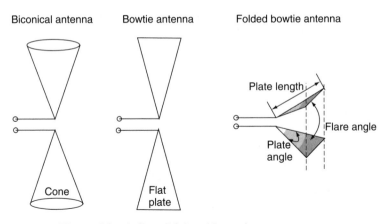

Figure 6.3   A few wideband bow-tie antenna types

Lu and Shi (1999) demonstrated that dipole structures should be favoured over monopole structures, because the size of the ground plane needed with the monopole may be not be practical. On the other hand, the problems in developing an ultra-broadband dipole radiator is in controlling the unbalanced currents in the outer feed cables. The currents are normally filtered by balun transformers or RF chokes. However transformers for UWB systems are difficult to design, and performance of ferrite RF chokes degrade above 1 GHz.

6.3.7   Loop Antennas

There have been several attempts to use loop antennas in UWB communications. In contrast to dipoles, the loop antennas radiate second derivatives of the incident electric field (Harmuth, 1978), which may be advantageous for some waveforms. Yarovoy *et al.* (2000) have presented a small loop antenna for UWB measurements. Such an antenna structure appears to be a good candidate for the portable antenna.

The so-called large current radiators can also be considered to be loop antennas. Their use in pulsed radiation was demonstrated by Pochain (1999).

6.3.8   Antenna Arrays

In UWB radar applications, linear and planar antenna arrays may be formed with very sparsely spaced elements. This enables economical, high resolution phased array antennas, with a beam which may be readily steered (Anderson *et al.*, 1991). The ratio of the wideband peak sidelobe level to the peak main lobe level is a function of the number of antenna elements rather than the element spacing. As a consequence, sparsely spaced wideband array antennas do not result in significant grating lobes. If the number of antenna elements is increased, the side-lobe levels may be reduced to almost arbitrarily low levels. In Lu and Shi (1997), an UWB omnidirectional monopole antenna is presented for the SPEAKeasy[t] system, where the bandwidth requirement for the base station antenna ranges from 30 to 500 MHz.

6.4 Simulation Techniques

The time and frequency domain are connected by the Fourier transform, which means that time domain antenna simulations can be performed using standard frequency domain electromagnetic simulators. However, conducting analysis in the frequency domain requires the calculation of far-field values for both amplitude and phase. This must be done over a very wide frequency range to be able accurately to extract the shape of the radiated pulse. For example, if transmission between two antennas is investigated (separated by 2 m), the calculation must be done over the range of approximately 10 MHz to 10 GHz, with a suitably small step size determined by phase resolution requirements, in order to preserve the pulse characteristics. In practice, this may mean a step size of 10–20 MHz, leading to a very large number of required calculation points. In this work it is assumed, that the calculations are performed using standard personal computers, which makes certain calculations very time consuming or, due to lack of memory, impossible to perform. The limitations are less severe, if modern supercomputers with parallel architectures are used instead.

In this section, the transition from the frequency domain to the time domain and visualization have been performed using Matlab[TM].

6.4.1 Finite-Element Method

Software tools based on use of the finite-element method (FEM), such as Ansoft HFSS, can be used to simulate antennas of arbitrary shape and material. However, FEM techniques are very slow. Furthermore, the FEM mesh evaluation technique works one single frequency at a time, implying that the mesh produced is only applicable for a limited bandwidth around the single frequency used in the calculation.

6.4.1.1 Method of Moments

Software tools based on the method of moments (MoM) are much faster than FEM tools. Most software tools, such as the Zeland IE3D and NEC-Win Pro from Nittany Scientific do not use adaptive meshing. Instead, the user may define the density of the grid. According to preliminary testing (see Section 6.5), moment-method-based tools are applicable for predicting the transient antenna behaviour. However, most MoM codes can only support infinite dielectric layers.

6.4.1.2 Finite-Difference Time-Domain

The finite-difference time-domain (FDTD) method and other time-domain methods are inherently applicable for studying the ultra-short pulses. Software packages such as

---

[t]USA military software defined radio system (Cook and Bonser, 1999).

CST Microwave Studio can be used to simulate antennas of arbitrary shape and material. This is demonstrated in Section 6.5. The transient solver of CST Microwave Studio, which is based on finite integration, is restricted to rather simple and small structures only. For this reason the calculations tend to become very slow for complex structures. The improved grid generation methods, such as sub-gridding, together with increased computing capacity makes time-domain methods the most promising for UWB simulations.

6.5 Simulation examples

This section presents simulation examples. Structures that are known to preserve and disperse the incident pulse waveform are presented in order to highlight the differences between good and bad UWB antennas. Some of the structures are designed to work at the different frequency range as defined by FCC (2002). However, each structure and waveform can be scaled to meet the required frequencies.

The first three examples are simple dipole structures, including a perfectly conducting dipole, a capacitively loaded dipole, and a resistively loaded dipole. The fourth example is a conical dipole. The fifth example is a log-periodic dipole array, which is known not to be suitable for pulsed radiation. All these five structures are simulated using the NEC-Win Pro software, which models the structures using wire segments. There are several non-commercial versions of NEC-code, which makes it easy to compare these results. Students are advised to make comparative calculations with these codes. Dipole structures are treated instead of monopoles, as their treatment is more reliable with the used software.

Finally, a well known UWB antenna, the TEM horn, is simulated using both MoM software Zeland IE3D and time-domain software CST Microwave Studio.

The incident waveform in the simulations is a differentiated Gaussian monopulse. The centre frequency of the waveform is chosen to be 3 GHz. The time and frequency domain representations are presented in Figures 6.4, 6.5 respectively.

6.5.1 Perfectly Conducting Dipole

As a first example, a thin dipole antenna was examined using the NEC-Win Pro simulation tool. The results of the calculations are presented in Figures 6.6, 6.7 and Table 6.2. The electric fields and the power in these figures are given in arbitrary units. The same applies in subsequent figures. The reflection coefficient and the radiated electric field into the bore sight direction, presented in Figure 6.6, indicate that the dipole resonates in half-wavelength intervals. This causes severe ringing in the antenna as seen in the time-dependent bore sight electric field and time-dependent power spectrum (Figure 6.6). The ringing effect is very clear from the power spectrum. This is demonstrated in Figure 6.7, which show magnification of the power spectrum with respect to time. Note that the negative power is due to spline interpolation applied to the calculated points. The spline interpolation has been done to resolve the positions of the smaller peaks.

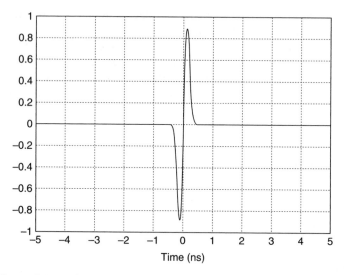

Figure 6.4 Incident voltage pulse used in time-domain simulations

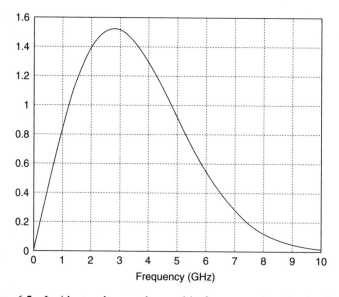

Figure 6.5 Incident voltage pulse used in frequency domain simulations

In Figure 6.7, peaks 1, 3, and 5 correspond to $(n + 1/2)\lambda$ resonances, while the peaks 2, 4, and 6 to $n\lambda$ resonances. The time differences from these peaks are presented in Table 6.2. The ringing is caused by reflections from the ends of the dipole. The electric field probe measuring the far field is located 2 metres from the antenna, which corresponds to a time delay of 6.7 ns.

Table 6.2 Time differences for peak positions
for power spectrum of thin dipole antenna

Peaks	Time delay (ns)
1–2	0.35
2–3	0.32
3–4	0.38
4–5	0.30
5–6	0.41
Half dipole distance (0.1 m)	0.33

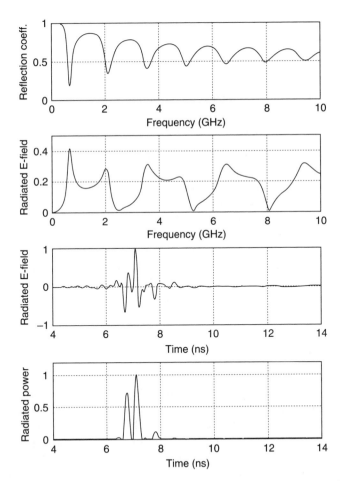

Figure 6.6 Reflection coefficient (uppermost), radiated electric field with respect to
frequency and time, and radiated power of a thin dipole antenna

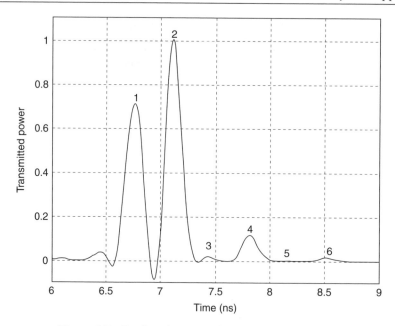

Figure 6.7   Radiated power of a thin dipole antenna

6.5.2   Capacitively Loaded Dipole

As a second example, a thin capacitively loaded dipole antenna was studied using the NEC-Win Pro software. The capacitive loading can be used to improve impedance matching of thin dipoles. For example, the impedance matching is improved compared with a typical dipole as shown in Section 6.5.1. However, the capacitances act as high-pass filters in the dipole antenna, which appears to increase the ringing effect here relative to that in the perfectly conducting dipole. This is illustrated in Figure 6.8. The electric field probe measuring the far field is located 1 metre from the antenna, which corresponds to a time delay of 3.3 ns.

The results for perfectly conducting and capacitively loaded dipole indicate that typical dipole structures are not optimal for UWB applications. The most straightforward method to improve the performance is to use resistive loading. This is demonstrated in the following Section 6.5.3.

6.5.3   Resistively Loaded Dipole

A resistively loaded Wu-King dipole (Maloney and Smith, 1993b) was also studied using the NEC-Win Pro software. The results, in Figure 6.9, indicate that the frequency domain gain and reflection coefficient are smooth. As a result of this, the pulse waveform is well preserved. The radiation efficiency of this antenna is, according to Montoya and Smith (1996), approximately 20%. In addition, the input impedance of the simulated antenna is as high as 660 Ω, which is impractical in realistic applications. On the other hand, the design clearly demonstrates the benefits of using resistive loading. In case high efficiency is not needed, the resistive loading is a straightforward method of improving the transient properties of small antennas.

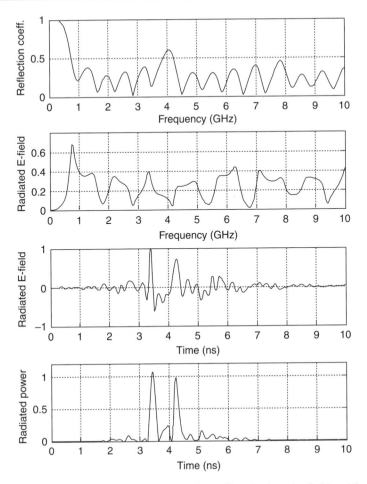

Figure 6.8 Reflection coefficient (uppermost), radiated electric field with respect to frequency and time, and radiated power of a capacitively loaded dipole

6.5.4 Conical Dipole

A conical dipole was studied using the NEC-Win Pro software as a example of a UWB omnidirectional antenna. The model was also used to demonstrate that the time-domain properties are generally direction dependent, as are the frequency-domain properties. The wire model of the antenna is presented in Figure 6.10. The results for this antenna in the bore sight direction are presented in Figure 6.11.

Figure 6.11 indicates that the frequency-domain radiation pattern of the bi-conical antenna in the bore sight direction is not smooth, but contains several deep minima. This appears to cause considerable ringing, although with much smaller amplitudes as in standard and capacitively loaded dipole (Sections 6.5.1 and 6.5.2). The radiation pattern is smoother in directions slightly above the bore sight plane. This is illustrated in Figure 6.12. The smooth gain also results in decreased ringing.

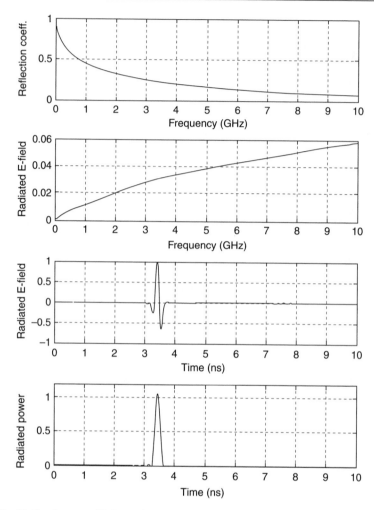

Figure 6.9 Reflection coefficient (uppermost), radiated electric field with respect to frequency and time, and radiated power for resistively loaded Wu-King dipole

It is possible to use resistive loading to decrease the ringing even further, as demonstrated by Maloney and Smith (1993a).

The time-domain behaviour of the bi-conical antenna wire model is in general better than that of the thin dipole. The conical dipole, with its variants such as the 2D bow-tie antenna, can thus be seen to be a good starting point for small UWB antenna designs.

6.5.5   Log-periodic Dipole Array

A log-periodic dipole array (LPDA, see Figure 6.13) was examined using the NEC-Win Pro software package as an example of a dispersive wideband antenna. The results are presented in Figure 6.14. The results indicate that the antenna examined has a very low

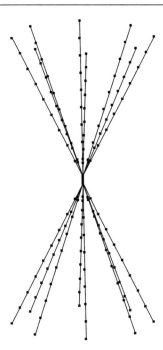

Figure 6.10   Wire-model of a bi-conical dipole

reflection coefficient and smooth gain. However, the LPDA does not have a constant phase centre, which causes severe dispersion of the pulse. Figure 6.14 shows that the high frequency components of the pulse are the first to arrive at the receiver as the feeding point of the antenna is at the high frequency part of the antenna. The lower frequencies are delayed, because the resonating longer dipoles are located behind the feeding point with respect to the bore sight direction. On the basis of the results, the LPDA is in general not suitable for UWB applications.

It is possible to decrease dispersion by using special feeding structures for each of the antenna elements of LPDA (Excell *et al.*, 1998).

6.5.6   TEM Horn

A TEM horn model was designed in order to compare the available software to measured data (see Section 6.6). Simulations were performed using a moment method based Zeland IE3D program package and a time domain software tool from CST Microwave Studio.

The Zeland IE3D results for antenna-to-antenna setup are presented in Figure 6.15. The results obtained using the CST Microwave Studio package are presented in Figure 6.16. The antennas in Figure 6.15 are located 2 metres from each other, whilst the antennas in Figure 6.16 are located 0.6 metres from each other. The longer inter-antenna distances in CST Microwave Studio are not practical due to computational intensity. Note the different frequency scale for the uppermost figure. In the CST simulations, the incident pulse differs slightly from the Gaussian monopulse. Otherwise the two simulators give very similar results.

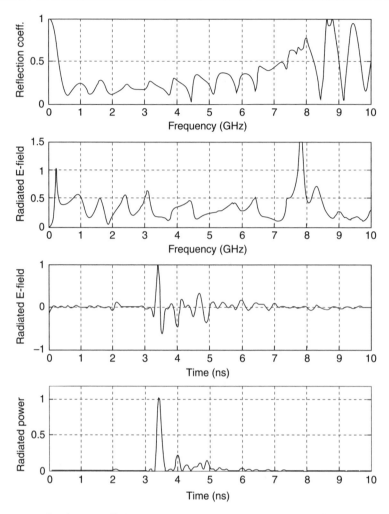

Figure 6.11 Reflection coefficient (uppermost), radiated electric field with respect to frequency and time, and radiated power for a bi-conical dipole (bore sight direction)

6.6 Measured examples

This section presents preliminary measurements for selected constructed prototypes. The measurement set-up required to measure antenna transient characterization is also briefly presented.

6.6.1 Measurement Techniques

6.6.1.1 Frequency-domain Measurements

Antenna-to-antenna time-domain responses can be evaluated from the corresponding frequency response. Both amplitude and phase of the transmission must be recorded.

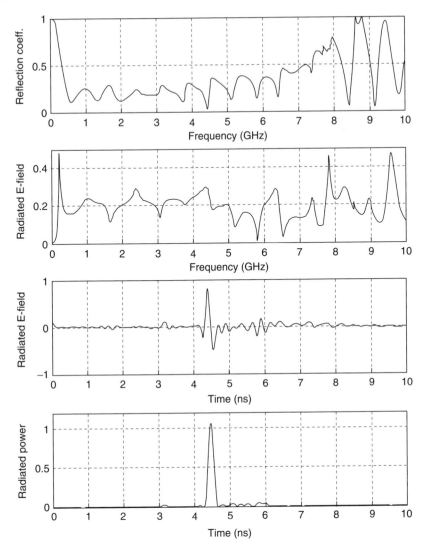

Figure 6.12 Reflection coefficient (uppermost), radiated electric field with respect to frequency and time, and radiated power for a bi-conical dipole (40 degrees above the bore sight plane).

This can be done using a standard vector network analyser (VNA). The VNA must be calibrated for a broad frequency span to avoid artificial distortion of the waveform due to aliasing. In a set of preliminary measurements, the VNA was calibrated to cover the frequency range of 50 MHz–20 GHz. This yields a maximum resolution of approximately 10 MHz in frequency, which according to the Nyquist rate formula, restricts the maximum measuring distance to about 5 metres. Longer distances can be handled by using multiple calibrations, or by narrowing the frequency range. Examples of measurements using VNA are presented below.

Figure 6.13 Wire model of log-periodic dipole array

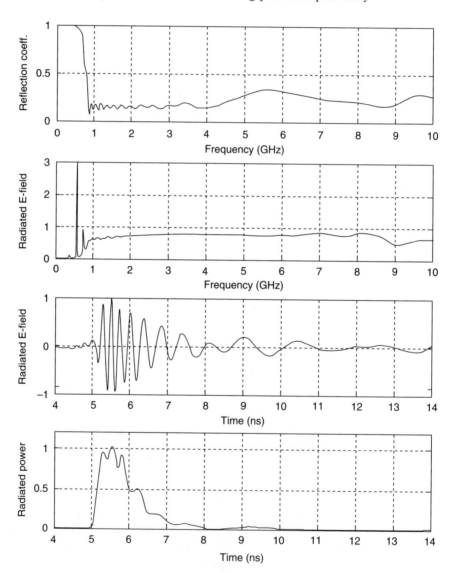

Figure 6.14 Reflection coefficient (uppermost), radiated electric field with respect to frequency and time, and radiated power for a log-periodic dipole array (bore sight direction)

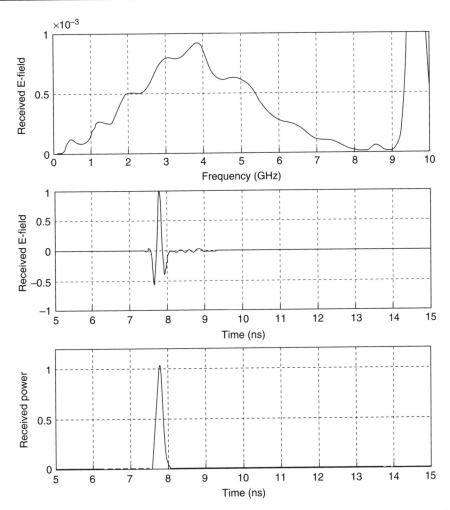

Figure 6.15 Received electric field with respect to frequency and time, and received time-domain power for TEM horn using the Zeland IE3D (bore sight direction)

6.6.1.2 Time-domain Measurements

A typical time domain antenna measurement consists of a pulse generator and a digital sampling oscilloscope. Most of the available pulse generators are based on step generators with very short rise and fall times. Pulses and impulses are generated from the step response using differentiation techniques. The pulses may also be shortened through the use of step recovery diodes. In practice, the authors have found that commercial pulse generators are expensive and difficult to embed in small designs. The lack of suitable commercial pulse generators has led to many designs being described in the open literature. For example, Lee *et al.* have presented small sub-nanosecond monocycle pulse generators (Lee and Nguyen, 2001a; Lee *et al.*, 2001c).

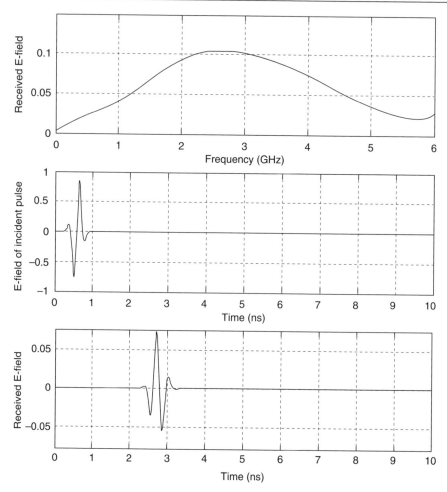

Figure 6.16 Received electric field with respect to frequency, incident waveform and received waveform for TEM horn, simulated using the CST Microwave Studio (bore sight direction)

A digital sampling oscilloscope, together with a sampling, head is a critical part of the measuring system.

6.6.2 TEM Horn

Two TEM horn prototypes were constructed and measured in order to demonstrate the use of the VNA in the time domain antenna characterization. The model, 'a bent bow-tie monopole' (see Section 6.5.6), is based on the simulated structure. For computational simplicity, the ground plate is infinite in the simulated model. The prototypes were constructed with different ground plane sizes in order to investigate the effects of a finite ground plane. The antennas are presented in Figure 6.17.

Figure 6.17   Constructed TEM horn with a small ground plane

The S_{11} of the TEM antenna using different prototypes and different simulation methods are presented in Figure 6.18. There is good agreement between the simulations and measurements for the antennas examined. In particular, the Zeland IE3D result very closely matched the experimental results. The antenna with the large ground plane is UWB according to our definition given in Table 6.1 — the VSWR is lower than 2 in the frequency range of 1.65 GHz–10.9 GHz, and lower than 3 in the frequency range of 0.93 GHz–18 GHz.

As a next step, the frequency-domain antenna-to-antenna measurements were performed. All measurements were performed with a VNA. Measured S_{21} values of amplitude and phase were post-processed with Matlab^{TM}, resulting in a mathematically calculated pulse shape. The measurements were done using two inter-antenna distances, 1 m and 1.5 m. In addition, a simulation in which a metal plate was placed between the antennas to cause reflections was performed with inter-antenna distance of 1 m. In the data analysis using MatlabTM, the antennas were 'excited' using the same Gaussian mono-pulse as in the Zeland IE3D simulations.

Figure 6.19 shows the simulated and measured waveform in the antenna-to-antenna set-up. The uppermost curve corresponds to the Zeland IE3D simulation, the second curve is from the CST Microwave Studio, and the three lowest curves are measured results (VNA measurements). The first measured result is obtained from the

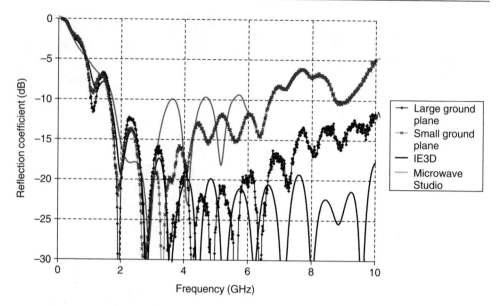

Figure 6.18   Measured and simulated S_{11} of a TEM horn

inter-antenna distance of 1 m, the second of 1.5 m. In the third measurement, the 1 m measurement is disturbed by a reflective plate.

The measured and simulated results indicate that the TEM horn is suitable for UWB operation. The results also indicate that the modern simulation software can be used reliably to estimate the transient behaviour of the UWB antennas as long as the antenna is modeled accurately.

6.6.3   Small Antennas

In mobile applications, the size of the antenna is often restricted. In this section two types of small antenna prototypes are designed and constructed.

The first of the prototypes is of modified bow-tie type. These antennas are later referred to as MBT (modified bow-tie). As compared with conventional bow-tie, the shape of the radiators is a half circle instead of triangle. This antenna uses a tapered-slot-type feed point, but acts as a bow-tie dipole at lower frequencies. This modification improves the impedance matching as compared to bow-tie antenna. The structure can be considered at a high frequency region as a pair of tapered slot antennas. Constructed antenna prototypes are presented in Figure 6.20. These antennas were manufactured on PCB. Measured and simulated S_{11} are presented in Figure 6.21. The agreement in this case is very good. The simulations were performed using CST Microwave Studio, as the use of the FI method makes it possible to model the finite PCB structure.

The second studied antenna type is a tapered slot antenna (TSA). The constructed TSA prototypes are shown in Figure 6.22. These antennas are more directive than the MBT antennas. The directivity makes these antennas less favourable for portable applications which need omni-directional radiation.

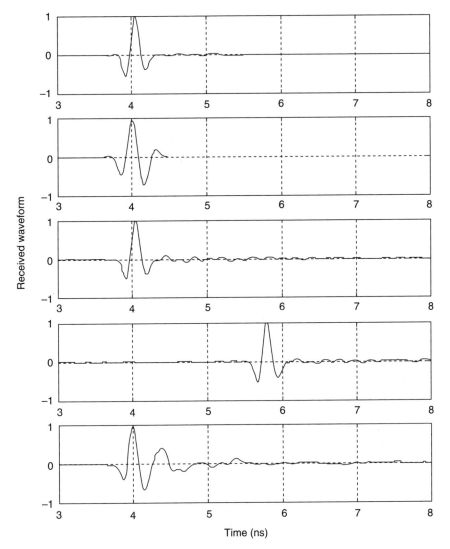

Figure 6.19 Simulated and measured waveform in the antenna-to-antenna set-up. See text for details and explanations

Figure 6.23 shows the simulated and measured waveform in the antenna-to-antenna set-up for MBT and TSA antennas. The results indicate that the ringing is not severe in these antenna types. The spurious pulse in MBT results at about 6 ns was found to be due to reflections from the antenna cables. The results indicate that these antennas can be considered a good starting point for mobile UWB antennas.

In order to demonstrate further that time-domain properties are generally direction dependent, the pulse shape of the MBT antenna was measured at different rotation angles (see Figure 6.24). Figure 6.25 shows the time-domain pulse as a function of the

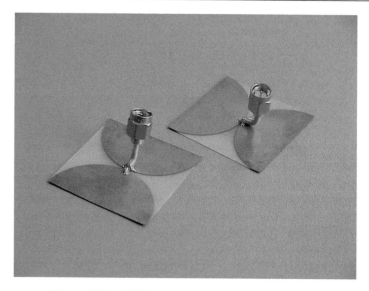

Figure 6.20   MBT antennas manufactured on PCB

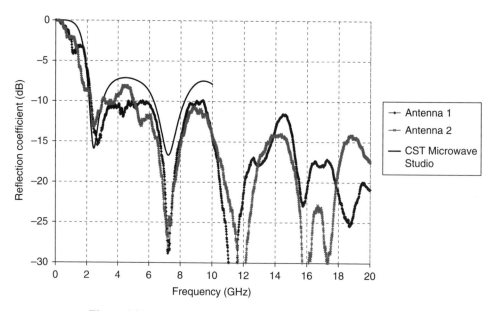

Figure 6.21   Measured and simulated S_{11} of MBT antennas

horizontal angle. The results indicate that the pulse shape varies as a function of the horizontal angle, but remains satisfactory at all angles.

In order for the pulse to be transmitted properly, the antennas must be of same polarization. Figure 6.26 shows the measurement set-up, in which the radiating antennas are of different polarizations. In this case, not only the amplitude of the

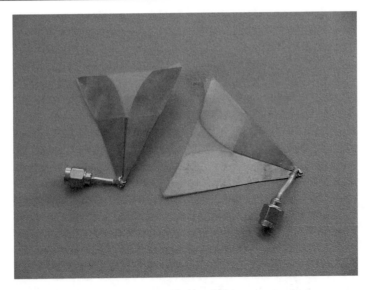

Figure 6.22   Tapered slot antenna prototypes

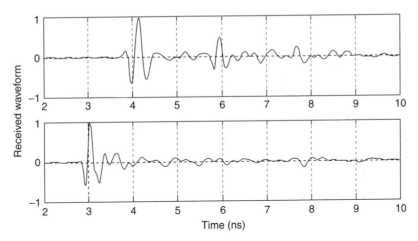

Figure 6.23   Measured waveform in the antenna-to-antenna setup for MBT (top) and TSA (below) antennas

pulse is decreased, but the pulse shape is heavily distorted. Figure 6.27 shows the time domain pulse as a function of the horizontal angle. Now the measured pulse shapes show a clear ringing effect. In mobile applications the orientation of the antenna cannot be guaranteed. For this reason, it is necessary to consider multiple polarizations, or alternatively circular polarization, to guarantee good pulse shape at the receiver.

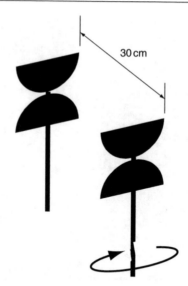

Figure 6.24 Measurement configuration for the MBT antennas. The horizontal angle is 0°

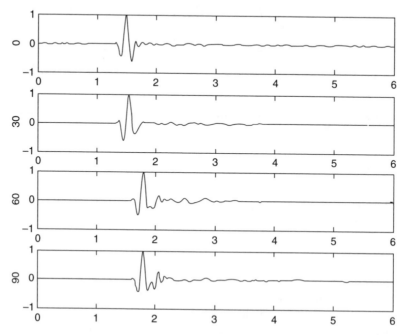

Figure 6.25 Time domain response at different rotation angles (0°–90°)

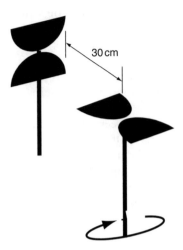

Figure 6.26 Cross-polarization measurement configuration for the MBT antennas. The horizontal angle is 0°

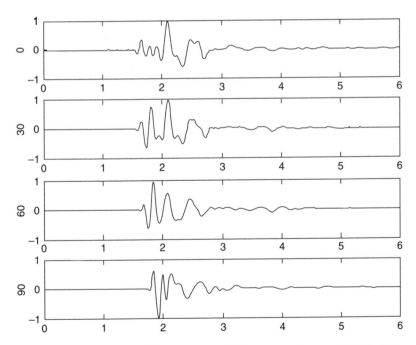

Figure 6.27 Time domain response at different rotation angles (0°–90°)

6.7 Conclusions

This chapter has examined some of the more common structures and analysis methods for UWB antennas. This has been done with the aid of simulations and prototype measurements. The results indicate that the design of UWB antennas can be done in a straightforward fashion using standard frequency domain simulators and measurements. The major design issues for UWB antenna are matching and ensuring constant gain, i.e. smooth amplitude and phase radiation patterns, across a very wide band of operation.

Structures for mobile antennas have been presented as well as various TEM horn structures for 'base station' antennas. Of the examples presented, the bow-tie, or a modification of it, is one of the candidates for mobile antennas due to its small size, simplicity, and the fact that it can be constructed using simple PCB techniques. The straightforward improvement of this antenna would be to use resistive loading in the end of radiators to minimize reflections.

7

Medium Access Control

Ulrico Celentano, Ian Oppermann

7.1 Introduction

UWB technology was originally seen purely as a physical layer technology, with no or little protocol to control the communication. It is now clear that Medium access control (MAC) features play a major role in UWB communication systems.

UWB holds enormous potential for wireless *ad-hoc* and peer-to-peer networks. One of the major potential advantages in impulse radio based systems is the ability to trade data rate for link distance by simply using more or less concatenated pulses to define a bit (see Chapter 3). Without dramatically changing the air interface, the data rate can be changed by orders of magnitude depending on the system requirements. This means, however, that high data rate (HDR) and low data rate (LDR) devices will need to coexist. The narrow time domain pulse also means that UWB offers the possibility for very high positioning accuracy. However, each device in the network must be 'heard' by a number of other devices in order to generate a position from a delay or signal angle-of-arrival estimate. These potential benefits, coupled with the fact that an individual low power UWB pulse is difficult to detect, offer some significant challenges for the multiple access MAC design.

UWB systems have been targeted at very HDR applications over short distances, such as USB replacement, as well as very LDR applications over longer distances, such as sensors and RF tags. Classes of LDR devices are expected to be very low complexity and very low cost, implying that the MAC will also need to be very low complexity. The potential proliferation of UWB devices means that a MAC must deal with a large number of issues related to coexistence and interoperation of different types of UWB devices with different capabilities. The complexity limitations of LDR devices may mean that very simple solutions are required. HDR devices, which are expected to be higher complexity, may have much more sophisticated solutions.

UWB Theory and Applications   Edited by I. Oppermann, M. Hämäläinen and J. Iinatti
© 2004 John Wiley & Sons, Ltd   ISBN: 0-470-86917-8

Research on UWB MAC is at an early stage, and no comprehensive solution for the unique difficulties for UWB systems have been proposed in the open literature. Existing or emerging standards that allow direct peer-to-peer communications include the IEEE 802.15.3 (IEEE, 2003) and ETSI HiperLAN Type 2, although only in the context of a centrally managed network. This chapter will examine some of the difficulties facing UWB MAC design, and look at the impact of some of the key physical layer parameters on MAC design. The chapter will also look at the emerging work of the IEEE working group 802.15.3a.

7.2 Multiple Access in UWB Systems

7.2.1 MAC Objectives

The very wide bandwidth of UWB systems means that many potential solutions exist to the issue of bandwidth usage. Devices may use all, or only a fraction of, the bandwidth available in the 3.1 to 10.6 GHz band. These devices will still be classed as UWB provided they use at least 500 MHz. Chapter 3 gave a summary of the major candidates for the physical layer signal structure of UWB systems, which include impulse radio, OFDM, multi-carrier and hybrid techniques. All of these possible techniques mean that different UWB devices may or may not be able to detect the presence of other devices. The main issues to be addressed by an UWB MAC include coexistence, interoperability and support for positioning/tracking.

7.2.1.1 Coexistence

The potential proliferation of UWB devices of widely varying data rates and complexities will require coexistence strategies to be developed.

Strategies for ignoring or working around other devices of the same or different type based on physical layer properties will reflect up to the MAC layer. Optimization of the UWB physical layer should lead to the highest efficiency, lowest BER, lowest complexity transceivers. The assumptions of the physical layer will however have implications on MAC issues such as initial search and acquisition process, channel access protocols, interference avoidance/minimization protocols, and power adaptation protocols. The quality of the achieved 'channel' will have implications on the link level, which may necessitate active searching by a device for better conditions, which is what happens with other radio systems.

7.2.1.2 Interoperability

The most common requirement of MAC protocols is to support inter-working with other devices of the same type. With the potentially wide range of device types, the MAC design challenge is to be able to ensure cooperation and information exchange between devices of different data rate, QoS class or complexity. In particular, emphasis must be placed on how low complexity LDR devices can successfully produce limited QoS networks with higher complexity, HDR devices.

7.2.1.3   Positioning/Tracking Support

Positioning is integrally linked to the MAC. This includes strategies for improving timing positioning accuracy and for exchanging timing information to produce positioning information.

It is possible for any single device to estimate the arrival time of a signal from another device based on its own time reference. This single data point in relative time needs to be combined with other measurements to produce a 3D position estimate relative to some system reference. Exchange of timing information requires cooperation between devices. Being able to locate all devices in a system presents a variation of the 'hidden node' problem. The problem is further complicated for positioning because multiple receivers need to detect the signal of each node to allow a position in three dimensions to be determined.

Tracking requires that each device is able to be sensed/measured at a suitable rate to allow reasonable update rate. This is relatively easy for a small number of devices, but difficult for an arbitrarily large number of devices. Information exchange between devices of timing and position estimates of neighbours (*ad hoc* modes) requires coordination, and calculation of position needs to be done somewhere (centralized or distributed) and the results fed to the information sink.

Finally, it is important to have the received signal as unencumbered by multiple access interference as possible in order to allow the best estimation of time of arrival. Every 3.3 ns error in delay estimation translates to a minimum 1 m extra error in position estimation.

All of these issues (information exchange, device sampling rate, node visibility, signal conditioning) require MAC support. These issues are significant obstacles to existing WLAN and other radio systems offering reliable positioning/tracking when added on to the MAC post-design.

7.2.2   Structure of the UWB Signal

As seen in Chapter 3, the impulse radio UWB signal is composed of a train of very short pulses. A typical pulse width T_p, is of the order of 0.1–0.5 ns. Figure 7.1 shows the structure of the basic UWB signal. Pulses are repeated with a (mean) pulse repetition time or frame time, T_f, $T_f \gg T_p$. The value of T_f may be hundreds or thousands of times the pulse width (Scholtz, 1993, Scholtz and Win, 1997). The reciprocal of T_f, R_f, is called pulse repetition frequency.

In the absence of coding or modulation, pulses are separated by T_f seconds. As seen in Chapter 3, pulses might be placed within the time frame period depending on the coding and modulation techniques applied.

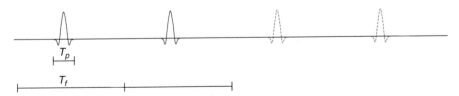

Figure 7.1   Structure of the basic UWB signal

In order to increase error resistance, repetition coding is often used. This means that a number of pulses, N, is transmitted for each bit. This may be seen as over-sampling with a factor of N when mapping data symbols to pulses. The bit period is given by

$$T_s = NT_f. \tag{7.1}$$

7.2.3 Modulation and Multiple Access

As seen in Chapter 3, common multiple access techniques adopted for pulse based UWB systems are TH and DS. Suitable modulation techniques include OOK (Foerster *et al.*, 2001) and especially PPM and PAM (Hämäläinen *et al.*, 2002). A given UWB communication system will be a combination of these techniques, leading to signals based on, for example, TH-PPM, TH-BPAM or DS-BPAM. TH-PPM is probably the most frequently adopted scheme and will be used in the following as an example for determining the resources available in a UWB system, resources that are to be managed by the medium access control.

To carry information, a modulation scheme is applied to a series of UWB pulses. If PPM modulation is adopted, the position of the pulse inside the slot window is chosen according to the data bit transmitted. A pulse is transmitted 'on time', or delayed by a certain quantity, ΔT, depending on whether a '1' or a '0' is transmitted. It is clear that all N pulses to which each bit is mapped are delayed by the same time shift.

In order to avoid collisions from different users, each pulse is placed inside a slot of duration T_c positioned inside the frame period T_f. The slot position is chosen according to a specific time hopping code which may be unique to each user (centralized system) or selected randomly by each user (uncoordinated system). N_h slots of duration T_c are available during a frame period T_f. The choice of N_h impacts the error performance and transceiver complexity. In an uncoordinated system, if N_h is small, the probability of user collisions is high. Conversely, if N_h is large, less time is available for reading the output and resetting the monocycle correlator (Scholtz, 1993; Win and scholtz, 1998a). For the same reason, the overall maximum time shift must be strictly less than the pulse repetition time T_f

$$N_h T_c < T_f. \tag{7.2}$$

In Figure 7.2, the solid-line user has time hopping code {2,1,3,2} and data {0,1}, whereas the dashed-line user has {2,3,1,1} and {1,1}. A third interfering signal is also shown in dot–dashed line.

For a TH-PPM system, the signal in the UWB channel in the multiuser case is given by:

$$\sum_{k=1}^{N_u} A_k \sum_{i=-\infty}^{+\infty} \sum_{j=0}^{N_s-1} g(t - iT_b - jT_f - c_j^{(k)} T_c - b_i^{(k)} \delta) \tag{7.3}$$

where $g(t)$ is the pulse waveform, A_k is the channel attenuation for user k, T_b is the bit period, $c_j^{(k)}$ is the time hopping code for frame j for user k, and $b_i^{(k)}$ is the data bit i for user k. The time hopping code chip is in the range from 0 to $N_h T_c$.

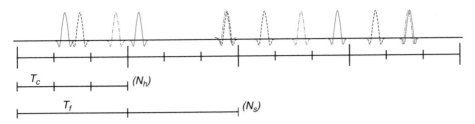

Figure 7.2 Example of TH-PPM. $N_h = 3$; $N_s = 2$

M-PAM is considered by Foerster *et al.*, (2001) and Hämäläinen *et al.*, (2002). The TH-BPAM signal for the kth user is given by:

$$s^{(k)}(t) = \sum_{i=-\infty}^{+\infty} \sum_{j=0}^{N_s-1} g(t - iT_b - jT_f - c_j^{(k)} T_c) d_i^{(k)} \tag{7.4}$$

where $d_i^{(k)}$ is the data sequence.

The DS-BPAM signal for the k-th user is given by:

$$s^{(k)}(t) = \sum_{i=-\infty}^{+\infty} \sum_{j=0}^{N_s-1} g(t - iT_b - jT_c) c_j^{(k)} d_i^{(k)} \tag{7.5}$$

where $d_i^{(k)}$ is the data sequence.

7.2.4 Multiuser System Capacity

The resistance against interference obtained with over-sampling, i.e., pulse repetition coding, described above, leads to a processing gain in linear scale expressed by

$$PG_1 = N \tag{7.6}$$

Another component of the processing gain is due to the duty cycle and can be called pulse processing gain (Foerster *et al.*, 2001)

$$PG_2 = \frac{T_f}{T_p} \tag{7.7}$$

The total processing gain is the product of these two contributions (Time Domain Corporation, 1998)

$$PG = PG_1 PG_2 \tag{7.8}$$

Since the duty cycle for DS-UWB is 100 %, therefore $T_p = T_f$. Hence, the processing gain in DS-UWB comes only from the pulse repetition.

The user's symbol rate (bit rate) R_s is given by (3.20).

$$R_s = \frac{1}{NT_f}.$$
(7.9)

Using (7.6) (7.7) and (7.8), for TH-UWB and DS-UWB, the user's symbol rate in equation (7.9) can be written

$$R_s = \frac{1}{T_p/PG} = \frac{1}{T_p/PG_1}\bigg|_{TH} = \frac{1}{T_p/PG}\bigg|_{DS}.$$
(7.10)

With the same bandwidth, data rate and pulse width, the low duty cycle of TH-UWB signals means that TH-UWB requires a lower N than DS-UWB to achieve the same processing gain. The advantage is therefore a higher allowable peak (pulse) power, and so improved interference compared with other systems.

The SNR for a TH-PPM system in the presence N_U of users, has been evaluated by Scholtz and Win (1997), assuming random hopping sequences and modelling the multiple access interference (MAI) as additive white Gaussian noise (AWGN), according to the standard Gaussian approximation. Under the previous assumptions, the bit SNR is given by (Scholtz and Win, 1997; Win 2000

$$\gamma^{-1}(N_U) = \gamma^{-1}(1) + M \sum_{k=2}^{N_u} \left(\frac{A_k}{A_1}\right)^2$$
(7.11)

where $\gamma(1)$ is the SNR in the single-link case and A_k is the signal attenuation along the path. The parameter M is given by (Scholtz and Win, 1997; Win 2000; Scholtz, 1993)

$$M^{-1} = \frac{Mm_p^2}{\sigma_a^2}$$
(7.12)

where m_p and σ_a^2 depend on the pulse waveform and the modulation parameter ΔT (Scholtz, 1993; Scholtz and Win, 1997) see (3.10) and (3.11). The expressions of m_p and σ_a^2 are (Scholtz, 1997)

$$m_p = \int_{-\infty}^{\infty} w_{rec}(t - \Delta T)V(t)dt$$
(7.13)

$$\sigma_a^2 = \frac{1}{T_f} \int_{-\infty}^{\infty} \left[\int_{-\infty}^{\infty} w_{rec}(t - s)V(t)dt\right]^2 ds$$
(7.14)

where $w_{rec}(t)$ is the received monocycle waveform and $V(t)$ is the embedded correlation template signal defined as

$$V(t; \delta) = w_{rec}(t) - w_{rec}(t - \delta).$$
(7.15)

Using (7.9), (7.12) can be rewritten as

$$M^{-1} = N \frac{m_p^2}{\sigma_a^2} = \frac{1}{R_s T_f} \frac{m_p^2}{\sigma_a^2}.$$

(7.16)

Using (7.12), (7.11) becomes (Scholtz and Win, 1997; Win, 2000)

$$\gamma^{-1}(N_U) = \gamma^{-1}(1) + \frac{\sigma_a^2}{N m_p^2} \sum_{k=2}^{N_u} \left(\frac{A_k}{A_1}\right)^2.$$

(7.17)

The ratio

$$\Delta P = \frac{\gamma(N)}{\gamma(1)}$$

(7.18)

is the increase in power required to keep performance constant with increasing number of users, N_U. It can be seen that $N_U(\Delta P)$ is monotonically increasing with ΔP. Under perfect power control (Scholtz and Win, 1997; Win and Scholtz, 1998a; Win, 2000

$$N_{U,max} = \lim_{\Delta P \to \infty} N(\Delta P) = \lfloor M^{-1} \gamma_{req}^{-1} \rfloor + 1$$

(7.19)

regardless of increase in transmit power.

The expression of the SNR depends on the pulse waveform correlation properties. One common type of pulse waveform is the Gaussian pulse and its derivatives. Different pulse waveforms have been considered in Chapter 3.

The transmit power is given by (Cuomo *et al.*, 1999)

$$P_t = \frac{E_w}{N_s T_c},$$

(7.20)

where E_w is the energy.

The expressions reported here are discussed in the following section.

7.3 Medium Access Control for Ultra-Wideband

The typical structure of cellular networks or access-point based networks is to have a central coordinating node which may be reached by each node in the network. In *ad-hoc* networks, the coordinator is dynamically selected among capable devices participating the network. In fully *ad-hoc* networks, or multi-hop networks, there is no coordinator at all and access is entirely distributed.

Regardless of the network topology, the role of the MAC is to control access of the shared medium, possibly achieving targeted goals such as high performance (quality of service provision), low energy consumption (energy efficiency), low cost (simplicity), and flexibility (for example with *ad-hoc* networking capabilities).

For all mobile applications, energy efficiency is an important consideration, and it is crucial in sensor networks. Energy efficiency affects the MAC design by requiring the protocol overhead due to signalling (connection set-up, changing configuration, etc.) to be minimized.

This section outlines the design foundations for a MAC suitable for operation using UWB technologies. The focus of the section is on the resource allocation problem (Figure 7.3).

The physical layer provides a bit stream to the upper layers, utilizing techniques and signals to optimize the usage of the available channel. At the physical layer, no distinction is made with respect to the significance of the bits carried. The role of the MAC is to arbitrate access to the resources made available from the physical layer.

The previous section presented expressions for key system parameters (such as the number of users, bit rates and achieved SNR) in terms of UWB parameters (such as pulse repetition rate, number of pulses per bit). The system parameters represent the resources made available by the physical layer. By adjusting these system parameters, it is possible to design a MAC tailored for UWB.

The wide range of values of the UWB parameters enables the design of highly adaptive software defined radios (di Sorte *et al.*, 2002) which may be matched to a wide range of service requirements (Cuomo et al.). In addition to focusing on the resources to be managed, possible constraints imposed by the UWB technology itself must be taken into account.

7.3.1 Constraints and Implications of UWB Technologies on MAC Design

Some qualities of UWB signals are unique and may be used to produce additional benefit. For example, the accurate ranging capabilities with UWB signals may be exploited by upper layers for location-aware services. Conversely, some aspects of UWB pose problems which must be solved by the MAC design. For example, using a

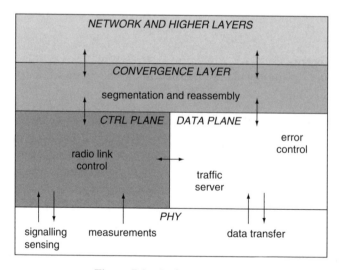

Figure 7.3 Reference model

carrier-less impulse radio system, it is cumbersome to implement the carrier sensing capability needed in popular approaches such as carrier-sense, multiple access with collision avoidance (CSMA/CA) MAC protocols.

As seen in Chapter 4, another aspect that affects MAC design is the relatively long synchronization and channel acquisition time in UWB systems. In (di Sorte *et al.*, 2002), the performance of the CSMA/CA protocol is evaluated for an UWB physical layer. CSMA/CA is used in a number of distributed MAC protocols and it is also adopted in the IEEE 802.15.3 MAC.

Figure 7.4 shows the packet traffic between two users in an UWB *ad-hoc* network. The approach assumes a very simple CSMA/CA protocol. Synchronization preambles must be sent at the beginning of each transmission burst. In order to exchange data with the CSMA/CA protocol, the receiver must first synchronize and decode the RTS (request to send) packet. The transmitter then needs to synchronize and decode the CTS (clear to send) packet, before starting transmission. In Figure 7.4, the UWB preambles are highlighted and denoted by 'p'. Processing times are neglected.

A long preamble is needed in order to achieve synchronization for both the RTS and CTS packets. A shorter preamble is possible for data transmission since synchronization may be assumed to be maintained after reception of the RTS packet. Further packets may be received with only fine corrections or tracking. Depending on the allowed length of the data packet, a longer preamble may be needed for the following ACK packet. Three preamble lengths (nominal, long, short) are proposed by Roberts (2003).

The time to achieve bit synchronization in UWB systems is typically high, of the order of few milliseconds (di Sorte *et al.*, 2002). Considering that the transmission time of a 10 000 bit packet on a 100 Mbit/s rate is only 0.1 milliseconds, it is easy to understand the impact of synchronization acquisition on CSMA/CA-based protocols. The efficiency loss due to acquisition time can be minimized by using very long packets. However this may impact performance in other ways.

When the effects of acquisition are taken into account, simulation results (di Sorte *et al.*, 2002) show that the performance of CSMA/CA using UWB is poorer than for narrowband and even wideband systems, in terms of delay, throughput and channel utilization.

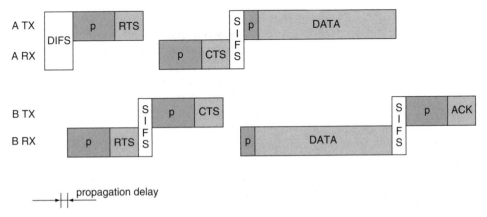

Figure 7.4 Messages between sender A and destination B using the CSMA/CA protocol

Acquisition preambles are typically sent with higher transmit power than data packets (Kolenchery *et al.*, 1998). This impacts both the interference level and the energy consumption in highly burst traffic. This effect must be taken into account when determining the efficiency of the system.

The adoption of CSMA/CA as a distributed protocol must be jointly evaluated with the performance of the underlying UWB physical layer. In general it may not be a suitable choice for an UWB MAC unless proper synchronization techniques are developed. One solution to this problem is the exploitation of the very low duty cycle of impulse radio. Synchronization can be maintained during silent periods by sending low power preambles for synchronization tracking (Kolenchery *et al.*, 1998). This approach is feasible only for communications between a single pair of nodes, which is not the case in peer-to-peer networks.

7.3.2 Resource Allocation in UWB Systems

The previous section gave an overview of multiple access in UWB. TH-PPM has been used to illustrate in detail how the UWB parameters influence system performance. Time-hopping multiple access is particularly interesting in *ad-hoc* networks since it does not require network synchronization, thus allowing asynchronous users (Win *et al.*, 1997b). This property is also a desirable feature in structured networks since it may relax the system requirements. The expressions describing system parameters, such as data rate given in Section 7.2.4 are valid for devices using a single time-hopping code. The results may be extended in a straightforward manner for the case of devices using multiple codes.

From the expressions given in Section 7.2.4, it is clear that there is a trade off between SNR (range) and data rate (throughput). For example, it can be seen that decreasing the frame time, T_f, implies a linear increase in the transmitted power (Foerster *et al.*, 2001). However, the pulse repetition frequency and the peak pulse power must be varied inversely to maintain constant average power. Hence, for a fixed T_f, to increase the receiver SNR, it is necessary to increase N and hence reduce R_s, trading off between SNR and data rate.

UWB systems allow devices to vary rate and performance simply by varying the number of pulses per bit. As seen above, N is related to the interference rejection capability of the user. This means that for loss-tolerant traffic, or for low-interference environments, N can be adjusted to trade between quality and bit rate. Multi-code allocation may be used to offer a wider range of achievable rates to a single terminal.

The UWB parameters listed in Table 7.1 affect MAC performance (di Benedetto 2001). These parameters can be adjusted to achieve the service requirement goals.

Cuomo *et al.* (1999) propose a resource allocation technique that optimizes system performance indices by varying the MAC parameters within given constraints. The authors start from the identification of UWB-specific MAC parameters to develop a distributed protocol, which supports guaranteed quality and best effort traffic. According to that protocol, a transmitter computes the allowed transmit power so that the interference to other users is kept below a threshold. This threshold is computed using the tolerable interference declared by each device in the network. The transmit rate is a constraint for the guaranteed quality, reservation-based traffic. The protocol determines whether a transmit power exists for a given user that satisfies the global requirements of

Table 7.1 Some UWB parameters, impact on data rate, R_s, and transmit power, P_t, and other comments

Parameter	Symbol	Comment
Pulse shape	—	Pulse shape can be used to control interference
Pulse duration	T_p	Impacts on level of external interference
Pulse repetition time	T_f	Impacts on coverage area (range) (Foerster *et al.*, 2001)
Chip time	T_c	Impacts on R_s, P_t (constrained to technology implementation)
Number of pulses per information bit	N	Impacts on R_s (simple to adjust); Impacts maximum power
Number of allocated TH codes	N_c	Impacts on data rate
Number of chips per pulse repetition frame	N_h	Impacts on R_s, P_t (see text; being an integer, no continuous values can be achieved: quantization effect) Impacts on UWB interference
Time shift Period of TH code	δ	Impacts on external interference
Transmission rate	N_p	Increasing N_p the power spectral density is controlled
Family of hopping codes Type of data modulation		Impacts on UWB interference

the network. For best-effort traffic, a solution always exists. The algorithm identifies suitable pairs of rate and power so that inequalities are satisfied.

7.4   IEEE 802.15.3 MAC

7.4.1   Introduction

The IEEE 802.15.3 working group on wireless personal area networks is developing a high data rate MAC capable of operation using UWB transmission. The current version (version 17) of the draft standard defines the physical layer and the MAC. The draft standard has been developed primarily to support 'classical' narrowband systems, and needs major changes to address UWB technology.

Within the IEEE working group is the study group (SG) 3a, which is the so-called alternate physical layer examining UWB solutions. At present, there are two main candidates for the IEEE UWB standard, neither of which has yet lead to concrete proposals at the MAC level. The first is a multiband approach (as described in Chapter 3), which is based on the flexible utilization of a number of UWB channels each of approximately 500 MHz. With this technique, the generated UWB signal is reshaped according to country-specific regulatory constraints and the presence of known narrowband interferers (such as WLAN devices). The second is based on impulse radio or pulse-based UWB techniques. In both cases, a flexible, dynamic MAC protocol must be developed to manage the system.

To date, little has been done for LDR UWB communication systems, and significant innovation is required to develop a low complexity MAC for these devices. A commonly shared view of future UWB systems is that LDR devices capable of achieving ultra-low power and ultra-low cost should include a processor-less MAC protocol, which may be operated using simple hardware finite state machines. Work within the IEEE working group 802.15.4 will attempt to address MAC issues for LDR devices.

7.4.2 Applications

The IEEE 802.15.3 standard is being developed for high speed applications including

- Video and audio distribution:
  -high speed DV transfer from a digital camcorder to a TV screen;
  -High definition (HD) MPEG2 (or better) between video players/gateways and multiple HD displays;
  -home theatre audio distribution;
  -PC to LCD projector;
  -interactive video gaming.
- High-speed data transfer:
  -MP3 players;
  -personal home storage;
  -printers and scanners;
  -digital still cameras.

To meet the above goals, the main characteristics of the IEEE 802.15.3 MAC are being defined as

- high rate WPAN with multimedia QoS provision:
  -short range (minimum 10 m, up to 70 m possible);
  -high data rates (currently up to 110 Mbit/s, to be increased by TG3a to 100–800 Mbit/s);
  -TDMA super-frame architecture.
- ad-hoc network with support for dynamic topology:
  -mobile devices may often join and leave the piconet;
  -short time to connect (< 1 s);
  -peer to peer connectivity.
- centralized and connection-oriented topology:
  -the coordinator maintains the network synchronization timing, performs admission control, assigns time for connection between devices, manages PS requests.
- flexible and robust:
  -dynamic channel selection;
  -transmit power control per link;
  -handover of the piconet coordinator (PNC) role among capable devices;
  -multiple power saving modes (support low power portable devices).
- authentication, encryption and integrity:
  -CCM authenticate-and-encrypt block cipher mode using AES-128 (IEEE, 2004);
  -support for upper layer authentication protocols (e.g., public key).

The main features are described in more detail in the following section.

7.4.3   Main Features

Although the IEEE 802.15.3 MAC is under consideration for use with UWB radio technologies, it has not been specifically designed for UWB. This section presents the IEEE MAC and considers its suitability as a MAC for a UWB system. The discussion suggests that the UWB MAC considerations described in Section 7.3.1 are only partially supported by the IEEE 802.15.3 MAC.

A wireless personal area network (WPAN) is a wireless *ad-hoc* data communications system, which allows a number of independent data devices to communicate with each other. A WPAN is distinguished from other types of data networks in that communications are normally confined to a person or object that typically covers about 10 metres in all directions, and envelops the person or a thing whether stationary or in motion.

The group of devices in the IEEE 802.15.3 MAC is referred to as a piconet as illustrated in Figure 7.5. The piconet is centrally managed by the coordinator of the network, referred to as the piconet controller. The PNC always provides the basic timing for the WPAN. Additionally the PNC manages the quality of service (QoS) requirements of the WPAN.

A device (DEV) willing to join a piconet first scans for an existing PNC. The presence of a PNC is detected by the reception of a beacon sent from the PNC on a periodic basis. If no PNC is found, and if the device is capable of the task, the new device becomes a PNC and starts a piconet itself. The PNC periodically sends network information via a beacon. Other devices that can receive the beacon may associate to the PNC. Associated devices can exchange data directly amongst themselves, i.e. without using the central node as a relay, but resources are managed centrally by the PNC. The role of PNC can be handed over to an elected device according to a pre-determined protocol.

A compliant physical layer may support more than one data rate. In each physical layer there is one mandatory base rate. In addition to the base rate, the physical layer may support rates that are both faster and slower than the base rate. A DEV will send a frame with a particular data rate to a destination DEV only when the destination DEV is known to support that rate.

The IEEE 802.15.3 MAC is based on a dynamic TDMA structure. The channel time is divided into periods referred to as 'super-frames' as shown in Figure 7.6. The

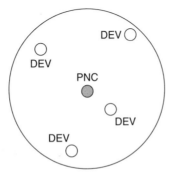

Figure 7.5   A piconet includes a piconet coordinator (PNC) and associated devices (DEVs)

structure and duration of the super-frames may vary from frame to frame. The super-frame is further divided into time slots and organized as below.

During the beacon time, the PNC broadcasts the beacon frame, which includes all the necessary network information for all devices of the piconet. The information includes allocation of the following time slots of the super-frame to a source-destination pair, based on the channel request commands sent by the DEVs. Either the source or the destination may be a broadcast address.

The contention access period (CAP) is used according to the distributed carrier sensing multiple access with collision avoidance (CSMA/CA) protocol. The CAP can be used for commands or asynchronous traffic.

The channel time allocation period (CTAP) is used for contention-less access of asynchronous and isochronous data streams. Channel time allocation (CTA) periods and management CTAs are allocated by the PNC. The CTAs allow devices to communicate without interference from other devices in the network.

The PNC may choose to allocate optional management CTAs (MCTAs), instead of the CAP, for sending commands to and from the PNC. Commands from DEVs to the PNC are sent in a so-called open MCTA, which has the broadcast address in the source field. The access mechanism in open MCTAs and in association MCTAs is slotted Aloha, whereas it is TDMA in other MCTAs. The association MCTAs are used by

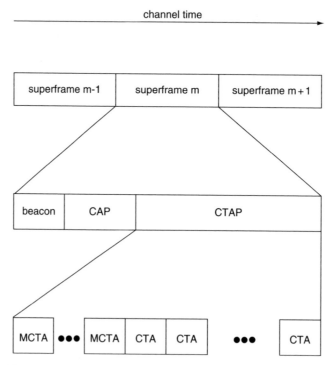

Figure 7.6   The super-frame structure includes contention and contention-less periods

DEVs wishing to join a piconet. Such DEVs use the MCTA to send their association messages.

CTAs are generally dynamic, meaning that their timing and position in the super-frame may change on a frame-by-frame basis. Conversely, pseudo-static CTAs are used so that their position and duration remain relatively constant. A third type, private CTAs, is reserved for uses other than communication, for example, to allow space for dependent piconets. The IEEE 802.15.3 MAC includes the possibility of creating so-called 'child piconets'. This is a feature that can be used to extend the coverage area by multiple hops, reducing the transmit power to lower the interference level. The frame structure for the child piconet is shown in Figure 7.7.

The PNC inserts guard times between consecutive CTAs (Figure 7.8). This guard time is defined so that the time separation between CTAs is always at least one SIFS (short inter-frame spacing). The required guard time depends on the maximum drift between DEVs' local time and the ideal time. It is therefore a function of the propagation delay and of the time elapsed since the previous synchronizing event, which is the start of the preamble of the beacon. The elapsed period from the synchronization event depends on the super-frame length. This means that clock drifts are a constraint for the super-frame length, and therefore for number and size of CTAs.

For data integrity, the IEEE draft standard includes three possible ACK policies. With *No-ACK*, the sending device does not expect any kind of guarantee that the frame is successfully received. With immediate *Imm-ACK*, the sending device requests and waits for an acknowledgment from the destination before sending the subsequent packet. Finally, the delayed *Dly-ACK* policy allows sending devices to send a number of frames while waiting for acknowledgement of the first transmitted frame. The buffer

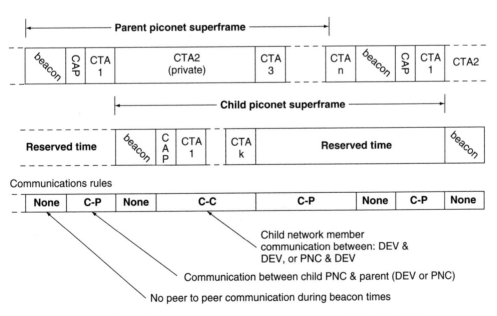

Figure 7.7   Child piconet frame structure (source IEEE 802.15.3)

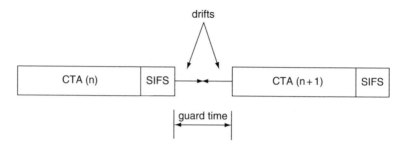

Figure 7.8 CTAs are separated by a guard time sufficient for compensating timing drifts

window size is negotiated between source and destination. The standard allows switching between Imm- and Dly-ACK

Figure 7.9 shows the reference model used in the IEEE draft standard. In the IEEE 802.15.3 MAC IEEE 2003, the resource reservation can be implemented exploiting the functionalities of MAC layer management entity (MLME) and device management entity (DME).

To reduce energy consumption, DEVs are allowed to enter a sleep mode, which may span a number of super-frames. The collision-free period (Channel Allocation Time Period) is TDMA based. The impact of synchronization acquisition time on CSMA/CA performance indicates that the usage of the CAP for asynchronous data transfer should be carefully considered.

7.4.3.1 UWB Considerations

As mentioned, the IEEE 802.15.3 MAC is not specifically designed for UWB. As a consequence, a number of issues must be considered when attempting to utilize UWB with the MAC.

Firstly, it is difficult for an UWB radio to detect when another DEV is transmitting. This may make much of the signalling difficult or impractical to implement. It may also mean that the CAP's may be zero length as multiple access is less of a problem than for conventional narrow-band systems.

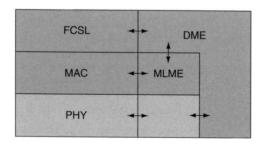

Figure 7.9 The reference model used in the draft standard

Secondly, the high data rates supported by UWB will allow a large number of devices to be in a piconet. Currently, the maximum number of unique DEV IDs in a piconet is 243, which will potentially limit the capacity of the UWB system.

Finally, high-data-rate UWB systems require efficient channel management to ensure high throughput. To maximize throughput, a physical-layer aware MAC should allow larger packet sizes and minimal spacing between frames from the same device type. The MAC protocols also need to be optimized to minimize the time required to associate with a piconet, quickly set up isochronous and asynchronous streams between DEVs, and monitor channel quality for optimum performance. This is something that the IEEE 802.15.3 MAC does not deal with particularly well for UWB systems.

7.5 Conclusions

At the time of writing, no standard exists for an UWB MAC. There is work in standardization bodies to adapt and possibly optimize and enhance existing MAC for use over UWB. Partly the reason is due to the inherent complications of using a physical-layer signal, which is difficult to detect and difficult to synchronize with for intended users. Efforts have been made to develop MACs that allow efficient operation for large numbers of devices. However, this work is still at an early stage.

An example of this is IEEE 802.15.3 MAC. Its critical points as a MAC for UWB have been outlined in this chapter. Whilst developed for high-speed, *ad-hoc*, network applications, the MAC has been targeted for UWB by interested groups. The suitability for UWB, either impulse based or for other physical layer UWB solutions, is yet to be proven.

Research on MAC for UWB is needed and is going on in industry research centres and in universities, although in research on MAC for UWB peculiarities of UWB signalling are rarely taken into account. As has been shown, limitations due to specific characteristics of UWB physical layer do exist and attention must be paid to these in MAC design. The potential of UWB systems to offer flexible data rates, large numbers of users and positioning services are critically dependent on the capabilities of the MAC.

8

Positioning

Keqen Yu, Ian Oppermann

8.1 Introduction

The very short time-domain pulses of UWB systems make them ideal candidates for combined communications and positioning. The duration of a pulse is inversely proportional to the bandwidth of the transmitted signal. If the time of arrival of a pulse is known with little uncertainty, then it is possible accurately to estimate the distance travelled by the pulse from the source. By combining the distance estimates at multiple receivers, it is possible to use simple triangulation techniques to estimate the position of the source.

For UWB systems with potential bandwidths of 7.5 GHz, the maximum time resolution of a pulse is of the order of 133 picoseconds. Therefore, when a pulse arrives, it is possible to know to within 133 picoseconds the 'time-of-flight' of the pulse. This time uncertainty corresponds to 4 cm spatial uncertainty. For more modest bandwidths of 500 MHz, the corresponding time resolution is 2 nanoseconds, which corresponds to a spatial uncertainty of approximately 60 cm. With any UWB signal, therefore, it is potentially possible to achieve sub-metre accuracy in positioning, provided the time-space uncertainties can be combined from multiple sources without significant loss. The corresponding location estimate will be subject to the cumulative errors of each of the distance estimates as well as any uncertainty or errors introduced by the positioning technique itself. Different techniques exist for determining position from time-of-arrival or time-of-flight estimates, and each technique has strengths and weaknesses.

The goal of this chapter is to explore positioning techniques as well as to look at some of the practical problems associated with positioning in UWB systems.

UWB Theory and Applications   Edited by I. Oppermann, M. Hämäläinen and J. Iinatti
© 2004 John Wiley & Sons, Ltd   ISBN: 0-470-86917-8

8.2 Positioning Techniques

8.2.1 Time-based Positioning

Positioning based on the use of radio signals has a long history. There are many position estimation techniques for various purposes under different scenarios. Signal strength, angle-of-arrival (AOA), time measurements (TOA, time-of-flight, and time difference of arrival (TDOA)) can all be exploited for the position estimation. Figure 8.1 shows a general positioning system configuration with base stations or sensors and the device to be located (tag). The example below is for a cellular system. However, the concept is generally applicable.

8.2.2 Overview of Position Estimation Techniques

The most straightforward way to estimate the position is directly solving a set of simultaneous equations (Fang, 1990) based on the TDOA measurements. Therefore, exact solutions can be obtained for two dimensional (2-D) positioning with three sensors/two TDOA measurements, and for three dimensional (3-D) positioning with four sensors. For an overdetermined system (with redundant sensors), Taylor series expansion may be used to produce a linearized, least-square solution iteratively to the position estimate (Torrieri, 1984). However, to maintain good convergence, the Taylor series method requires quite an accurate initial position estimate, which is often difficult to obtain in some practical applications.

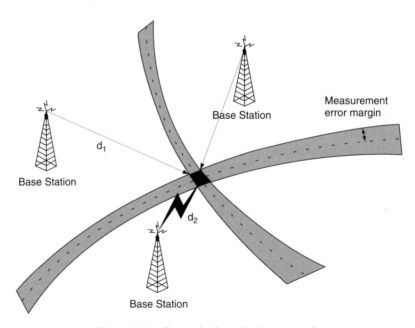

Figure 8.1   General triangulation example

To avoid the convergence problem, several different approaches have been proposed such as the spherical interpolation (Friedlander, 1987; Schau and Robinson, 1987; Smith and Abel, 1987) and the double maximum likelihood (ML) method (Chan and Ho, 1994). The hyperbolic positioning methods (except for the Taylor series method) share one common drawback of producing multiple solutions. They also require knowledge of the variance and the distribution of the TOA estimation error (except for the direct method). A different method for positioning is the use of non-linear optimization theory. Gauss–Newton method, Levenberg–Marquardt method, and quasi-Newton method, including the DFP formula (Fletcher and Powell, 1963) and the BFGS formula (Broyden, 1970) can all be employed for position estimation iteratively. The DFP algorithm has been used e.g. in the UWB precision assets location system developed by Multi-spectral Solution, Inc.

For practical applications, the position-estimation algorithm should be robust and easy to implement. To achieve this, we choose the direct calculation method and the non-linear optimization algorithm for further investigation. These two methods do not require knowledge of the variance or distribution of the TOA estimation error. They also do not require an accurate initial position estimate.

8.2.3 Direct Calculation Method

In the Cartesian system, the range (distance) between sensor i and the tag is given by

$$\sqrt{(x - x_i)^2 + (y - y_i)^2 + (z - z_i)^2} = c(t_i - t_0) \quad i = 1, 2, 3, 4 \tag{8.1}$$

where (x, y, z) and (x_i, y_i, z_i) are the coordinates of the tag and the sensor respectively, c is the speed of light, t_i is the signal TOA at sensor i, and t_0 is the unknown transmit time at the tag/device. In the development of the expressions, we ignore the difference between the true and the measured TOAs for simplicity. Squaring both sides of (8.1) gives

$$(x - x_i)^2 + (y - y_i)^2 + (z - z_i)^2 = c^2(t_i - t_0)^2 \quad i = 1, 2, 3, 4. \tag{8.2}$$

Subtracting (8.2) for $i = 1$ from (8.2) for $i = 2, 3, 4$ produces

$$ct_0 = \frac{1}{2}c(t_1 - t_i) + \frac{1}{2c(t_1 - t_i)}(\beta_{i1} - 2x_{i1}x - 2y_{i1}y - 2z_{i1}z) \quad i = 2, 3, 4 \tag{8.3}$$

where

$$x_{i1} = x_i - x_1$$
$$y_{i1} = y_i - y_1$$
$$z_{i1} = z_i - z_1$$
$$\beta_{i1} = x_i^2 + y_i^2 + z_i^2 - (x_1^2 + y_1^2 + z_1^2).$$

Define the TDOA between sensors i and j as

$$\Delta t_{ij} = t_i - t_j$$

Eliminating t_0 from (8.3) yields

$$a_1 x + b_1 y + c_1 z = g_1 \qquad (8.4)$$

where

$$a_1 = \Delta t_{12} x_{31} - \Delta t_{13} x_{21}$$
$$b_1 = \Delta t_{12} y_{31} - \Delta t_{13} y_{21}$$
$$c_1 = \Delta t_{12} z_{31} - \Delta t_{13} z_{21}$$
$$g_1 = \frac{1}{2}(c^2 \Delta t_{12} \Delta t_{13} \Delta t_{32} + \Delta t_{12} \beta_{31} - \Delta t_{13} \beta_{21}).$$

and

$$a_2 x + b_2 y + c_2 z = g_2 \qquad (8.5)$$

where

$$a_2 = \Delta t_{12} x_{41} - \Delta t_{14} x_{21}$$
$$b_2 = \Delta t_{12} y_{41} - \Delta t_{14} y_{21}$$
$$c_2 = \Delta t_{12} z_{41} - \Delta t_{14} z_{21}$$
$$g_2 = \frac{1}{2}(c^2 \Delta t_{12} \Delta t_{14} \Delta t_{42} + \Delta t_{12} \beta_{41} - \Delta t_{14} \beta_{21}).$$

Combining (8.4) and (8.5) yields

$$x = Az + B \qquad (8.6)$$

where

$$A = \frac{b_1 c_2 - b_2 c_1}{a_1 b_2 - a_2 b_1}$$
$$B = \frac{b_2 g_1 - b_1 g_2}{a_1 b_2 - a_2 b_1}$$

and

$$y = Cz + D \qquad (8.7)$$

where

$$C = \frac{a_2 c_1 - a_1 c_2}{a_1 b_2 - a_2 b_1}$$
$$D = \frac{a_1 g_2 - a_2 g_1}{a_1 b_2 - a_2 b_1}.$$

Then, substitution of (8.6) and (8.7) back into (8.3) with $i = 2$ produces

$$c(t_1 - t_0) = Ez + F \tag{8.8}$$

where

$$E = \frac{1}{c\Delta t_{12}} (x_{21} A + y_{21} C + z_{21})$$

$$F = \frac{c\Delta t_{12}}{2} + \frac{1}{2c\Delta t_{12}} (2(x_{21} B + y_{21} D) - \beta_{21}).$$

Substituting (8.6), (8.7) and (8.8) back into (8.1) for $i = 1$ followed by squaring yields

$$Gz^2 + Hz + I = 0 \tag{8.9}$$

where

$$G = A^2 + C^2 - E^2 + 1$$
$$H = 2(A(B - x_1) + C(D - y_1) - z_1 - EF)$$
$$I = (B - x_1)^2 + (D - y_1)^2 + z_1^2 - F^2.$$

The two solutions to (8.9) are

$$z = -\frac{H}{2G} \pm \sqrt{\left(\frac{H}{2G}\right)^2 - \frac{I}{G}}. \tag{8.10}$$

The two estimated z values (if both are reasonable) are then substituted back into (8.6) and (8.7) to produce the coordinates x and y, respectively. However, there is only one desirable solution. We can dismiss one of the solutions if it has no physical meaning or it is beyond the monitored area. If both solutions are reasonable and they are very close, we may choose the average as the position estimate. Otherwise, an ambiguity occurs. Other cases of no acceptable results include two complex solutions, or both solutions are beyond the monitored area. To increase the probability of the existence of one reasonable position estimate, we may add a fifth sensor. Then we have five different combinations, producing five different results.

In practice, there may be more than five sensors. In this case, we may choose the five sensors with the highest received signal powers. This simple method is particularly suitable when the noise level is similar in the received signals. When a sequence of measurements is available at each sensor, accuracy may be further improved by first processing the sequence of measurements such as averaging.

8.2.4   Optimization Based Methods

There are many schemes and techniques in non-linear optimization. In this section, we are interested in examining several practical optimization methods and applying them to our three-dimensional position estimation.

8.2.4.1   Objective Function

An objective function is normally required for any optimization algorithms. Since the ultimate aim of positioning is to obtain an accurate position estimate, it is natural to define the objective function as the sum of the squared range errors of all sensors

$$F(x, y, z, t_0) = \frac{1}{2} \sum_{i=1}^{N} f_i^2(x, y, z, t_0).$$

where N is the number of the active sensors/base stations, (x,y,z) is the unknown position coordinates, t_0 is the unknown transmit time (dummy variable), and

$$f_i(x, y, z, t_0) = \sqrt{(x - x_i)^2 + (y - y_i)^2 + (z - z_i^2)} - c(t_i - t_0).$$

Here t_i is the estimated TOA at the ith sensor and c is the speed of light. The optimization purpose is to minimize this objective function to produce the optimal position estimate. For notational simplicity, we define

$$\mathbf{p} = (x, y, z, t_0)^T$$
$$\mathbf{f}(\mathbf{p}) = (f_1(\mathbf{p}),\ f_2(\mathbf{p}), \ldots,\ f_N(\mathbf{p}))^T.$$

Then the objective function becomes

$$f(\mathbf{p}) = \frac{1}{2} ||\mathbf{f}(\mathbf{p})||^2.$$

8.2.4.2   Gauss–Newton Method

Expanding the objective function in the Taylor series at the current point $\mathbf{P}_k$ and taking the first three terms, we have:

$$F(\mathbf{p}_k + \mathbf{s}_k) \approx F(\mathbf{p}_k) + \mathbf{g}_k^T \mathbf{s}_k + \frac{1}{2} \mathbf{s}_k^T \mathbf{G}(\mathbf{p}_k) \mathbf{s}_k \qquad (8.11)$$

where S_k is the directional vector (or increment vector) to be determined, g_k is a vector of the first partial derivatives (also called gradient) of the objective function at P_k

$$g_k = \nabla f(x, y, z, t_0)|_{\mathbf{p}=\mathbf{p}_k}$$
$$= \left[\frac{\partial f}{\partial x}|_{\mathbf{p}=\mathbf{p}_k}, \frac{\partial f}{\partial y}|_{\mathbf{p}=\mathbf{p}_k}, \frac{\partial f}{\partial z}|_{\mathbf{p}=\mathbf{p}_k}, \frac{\partial f}{\partial t_0}|_{\mathbf{p}=\mathbf{p}_k}\right]^T$$

and $G(P_k)$ is the Hessian of the objective function. Minimization of the right-hand side of (8.11) yields

$$G(\mathbf{p})_k s_k = -g_k. \tag{8.12}$$

The minimization in which S_k is defined by (8.12) is termed Newton's method. To avoid the calculation of the second order information in the Hessian, a simplified expression can be approached from (8.12), resulting in

$$J_k^T J_k s_k = -J_k^T \mathbf{f}(\mathbf{p}_k) \tag{8.13}$$

where J_k is the Jacobian matrix of $\mathbf{f}(P)$ at P_k. This is called the Gauss–Newton method. When J_k is full rank, which is the usual case of an over-determined system, we have the linear least-squares solution

$$s_k = -(J_k^T J_k)^{-1} J_k^T \mathbf{f}(\mathbf{p})_k. \tag{8.14}$$

The Gauss–Newton method may get into trouble when the second-order information in the Hessian is not trivial. A method that overcomes this problem is the Levenberg–Marquardt method. The Levenberg–Marquardt search direction is defined as the solution of the equations

$$(J_k^T J_k + \lambda \mathbf{I}) s_k = -J_k^T \mathbf{f}(\mathbf{p}_k) \tag{8.15}$$

where λ is a non-negative scalar that controls both the magnitude and direction of S_k.

8.2.4.3 Quasi-Newton Method

This type of method is like Newton's method. The Hessian matrix $G(P_k)$ in (8.12) is now approximated by a symmetric positive definite matrix B_k, which is updated from iteration to iteration. At the kth iteration, set

$$S_k = -B_k g_k. \tag{8.16}$$

Using line search along S_k to produce

$$\mathbf{p}_{k+1} = \mathbf{p}_k + \alpha s_k \tag{8.17}$$

where α is the step size. Then updating $\mathbf{B}_k$ yields $\mathbf{B}_{k+1}$. The initial matrix $\mathbf{B}_1$ can be any positive definite matrix. It is usually set to be an identity matrix in the absence of any better estimate. There exist different ways to update $\mathbf{B}_k$. One well-known updating formula is the DFP (Davidon–Fletcher–Powell) formula, in which $\mathbf{B}_k$ is updated according to

$$\mathbf{B}_{k+1} = \mathbf{B}_k + \frac{\mathbf{h}_k \mathbf{h}_k^T}{\mathbf{h}_k^T \mathbf{q}_k} - \frac{\mathbf{B}_k \mathbf{q}_k \mathbf{q}_k^T \mathbf{B}_k}{\mathbf{q}_k^T \mathbf{B}_k \mathbf{q}_k} \tag{8.18}$$

where

$$\mathbf{h}_k = \mathbf{p}_{k+1} - \mathbf{p}_k$$
$$\mathbf{q}_k = \mathbf{g}_{k+1} - \mathbf{g}_k.$$

The BFGS quasi-Newton algorithm was also considered. This is significantly more complicated than the DFP algorithm. Preliminary simulation results demonstrated that the performance of the BFGS algorithm is not better than the DFP algorithm. As a result, only results for the DFP algorithm are presented.

To start the iteration for any of the above mentioned algorithms, the initial position coordinates and the initial transmit time are required. In the absence of any better estimate, the initial estimated values of the position coordinates may be chosen to be the mean position of all the active sensors or the area being monitored. The initial estimated transmit time may be chosen to be some time point earlier than the earliest receive time, which will depend on the dimension of the monitored area. Both the step size α and the first derivatives $\mathbf{g}$ (the gradient) are updated during each iteration. The performance of the above optimization-based methods could be improved if a good initial position estimate is available. This may be achieved by exploiting the results from the direct calculation method or other non-iterative methods.

8.2.5   Simulation Results

A simulation model was developed to examine the relative performance of the different positioning techniques. The tool models a closed region with a number of sensors and examines the accuracy with which a device in the area may be positioned in 3-D. The monitored area has been assumed to have a dimension of $30\,\text{m} \times 40\,\text{m} \times 5\,\text{m}$. The positions of the sensors (delay estimation points) and the tag (device of interest) are randomly generated to obtain average performance. The performance evaluation is first performed by assuming that the TOA measurement error is an i.i.d. random-variable Gaussian distributed with zero mean and variance s^2. At each value of s examined, 1000 simulation runs are conducted with random positions for the sensors and the tag at each run. The performance is then averaged. It is known that the location of the base stations can have significant impact on the positioning performance. The case of one particular location of the base stations is also examined. This location is randomly chosen from 50 locations of the base stations which produce the best results.

The performance is evaluated in terms of the root-mean-square (RMS) error and the failure rate. The RMS error is calculated according to

$$\sqrt{\frac{1}{3N_p N_s} \sum_{i=1}^{N_p} \sum_{j=1}^{N_s} [(x^{(i)} - \hat{x}^{(ij)})^2 + (y^{(i)} - \hat{y}^{(ij)})^2 + (z^{(i)} - \hat{z}^{(ij)})^2]}$$

where N_p is the number of different position combinations of the sensors and the tag and N_s is the number of TOA samples at each s for each position combination. $(x^{(i)}, y^{(i)}, z^{(i)})$ and $(\hat{x}^{(ij)}, \hat{y}^{(ij)}, \hat{z}^{(ij)})$ are the true and estimated position coordinates of the tag/device of interest, respectively. The failure rate accounts for the cases where there is no solution or the solution is unreasonable. In the direct calculation, the cases include: the solutions are beyond the monitored area; both solutions are complex valued; the two solutions are reasonable but not close to each other. With the optimization-based methods, it accounts for the situations where the algorithm does not converge to a solution, the maximum number of function evaluations or iterations is exceeded, or the results are beyond the area.

Figures 8.2 and 8.3 show the corresponding root-mean-square error of the three coordinates and failure rate for a 4/5 sensor system with Gaussian distributed TOA estimation error. DFPran and DFPsel denote the average RMS error of the x, y and z coordinate estimation using the DFP algorithm with five sensors with randomly chosen, and selectively chosen sensor positions respectively. LMran and LMsel are the

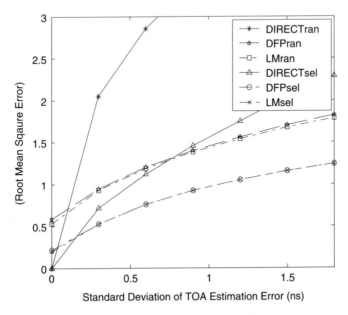

Figure 8.2   Root-mean-square error of position estimation

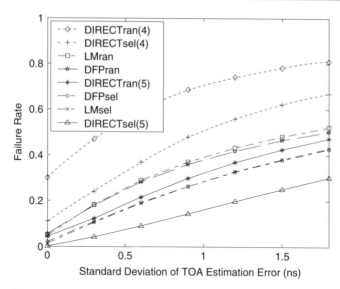

Figure 8.3 Position estimation failure rate for four- and five-sensor system

corresponding results using the Levenberg–Marquardt method, while DIRECTran and DIRECTsel are the corresponding results with the direct calculation using five sensors. The RMS error of the direct calculation method using four sensors is nearly the same as the case of five sensors, so it is not plotted. Clearly, for the direction-calculation method, the failure rate decreases dramatically with five sensors compared with the case of four sensors, although the accuracy is nearly the same. Of course, the failure rate improvement is achieved at the cost of increased computation complexity and system complexity, due to the addition of one extra sensor. The relatively large position error for small TOA estimation error in the DFP algorithm and the Levenberg–Marquardt method may come from the very crude initial position estimate. It seems the direction calculation method is suitable for quite small TOA estimation error, while the quasi-Newton algorithm and the Levenberg–Marquardt method are suitable for relatively large TOA estimation error. The Gauss–Newton method works very poorly so that the corresponding results are not presented.

Figures 8.4 and 8.5 show the results for the position estimation combined with delay estimation. A two-step technique is employed to speed up synchronization. Each data bit in the preamble assigned for synchronization consists of two separated pulse sequences, one of which is not coded and the other is coded by a m-sequence, and a large non-pulse region. The first step is to provide a rough timing message using the uncoded pulse sequence based on a moving average scheme with a window equal to the duration of the uncoded pulse sequence. Once the first step is completed, the approximate location of the coded pulses is available and the code acquisition proceeds to achieve fine synchronization. We choose the direct-sequence spreading with spreading gain 31 (i.e. 31 chips per data bit). The pulse width is 0.4 nanoseconds and the duty cycle is 1/14. The sampling rate is 6 GHz.

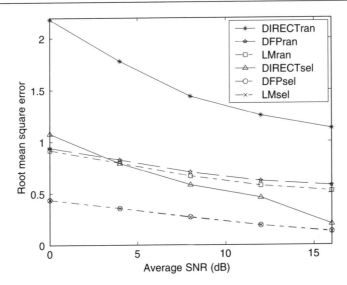

Figure 8.4 Root-mean-square error of position estimation using estimated TOA

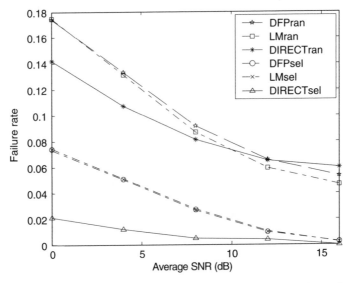

Figure 8.5 Position estimation failure rate for five-sensor system using estimated TOA

8.3 Delay Estimation Techniques

Conventional delay estimation uses a variety of techniques to improve the estimation of the position of a multipath component in a received signal. The delay estimation approaches typically attempt to detect the presence of 'L' multipath components. Figure 8.6 shows an example impulse response measurement for LOS conditions. As

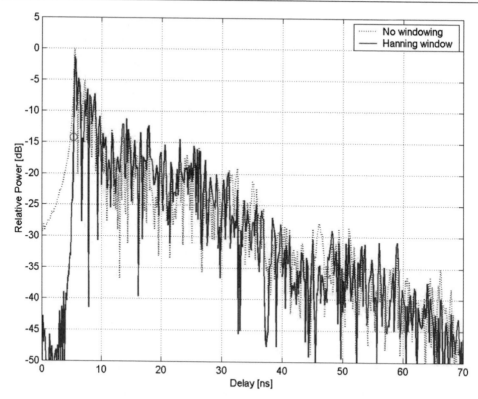

Figure 8.6 Measured channel impulse response (with and without window filtering)

was seen in Chapter 2, the UWB channel is very multipath rich, which makes detection of the direct component difficult.

Delay estimation approaches used for UWB are typically very simple due to the multipath-rich nature of the channel. A simple threshold detector may be used to detect the presence of the signal. However, much of the advantage of the small time uncertainty is lost with such a simple technique. The rich channel requires significant signal processing to produce an accurate estimate of time of arrival of the direct component. Several delay estimation techniques will be described here for completeness.

8.3.1 General Approaches

The delay estimation approaches generally used are interference/inter-path cancellation based on recognizing the shape of the band limited transmitted pulse such as described in (Moddemeijer, 1991). This approach is robust, but does not lead to enormous improvements in initial delay position estimation.

Subspace techniques such as those presented by Jakobsson *et al.*, (1998) are extremely complex, requiring the generation of several correlation matrices and their inverses, and ultimately performing a large number of matrix multiplications to achieve a delay estimate. They also tend not to work well in static or slowly moving channels (Latva-aho, 1998).

For example, eigenvector decomposition is a form of subspace technique (Manabe and Takai, 1992). This delay estimation approach requires calculation of the eigenvectors of the channel correlation matrix. Again, it is very complex.

8.3.2   Inter-path Cancellation

When using short spreading sequences, the autocorrelation side-lobes have non-zero values which contribute to each 'apparent' multipath component. As a consequence, all multipath components within at least one symbol period of the first 'detected' multipath will need to be considered.

The technique requires interpolation between samples and good knowledge of the shape of the band-limited single path signal.

The technique may be described as:

(1) identify shape of band limited autocorrelation (template);
(2) perform autocorrelation of received signal (this is the channel impulse response);
(3) estimate AWGN power;
(4) over-sample/interpolate to desired temporal resolution;

Loop

(5) identify largest multipath component in received signal;
(6) scale main lobe of template to fit (5);
(7) subtract (6) from (5);
(8) repeat **Loop** until all paths identified in area of interest or signal are below noise.

Once an estimate of the AWGN had been removed, the individual 'resolvable' multipath components are identified. This is achieved by selective identification and removal of the largest multipath components. Identification of large multipath components uses the fact that, given the finite bandwidth, each multipath has a characteristic shape. A template with this shape is fitted to the channel magnitude, and each large multipath component identified is subtracted from the impulse response magnitude. With some tolerance for numerical and measurement accuracy, anything remaining after the large multipath component (MPC) is removed must be due to other multipath components. The same treatment is therefore applied to the magnitude of the residue of the impulse response.

This technique requires the centre of the multipath component to be estimated from the samples that indicate the position of a given multipath component. Interpolated 'impulse templates', with resolutions higher than the received signal sampling rate, may be used to improve the accuracy of this approach. These are then correlated with the received magnitude until a maximum value is reached. The scaled template is then subtracted from the received signal.

Figure 8.7 shows an example of an impulse response which has had noise removed and multipath components detected. If two multipath components are separated by one sample, and less than approximately 8 dB, then they will be resolved. The power requirement comes from the error sensitivity mentioned earlier. Once an identified impulse is removed, a margin of error is allowed for to account for system noise and

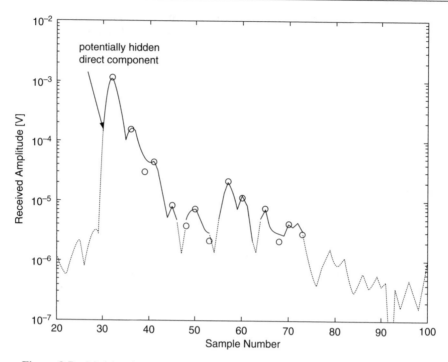

Figure 8.7   Multipath components identified in a measured impulse response

finite resolution of the measurements. If the residual signal exceeds this margin of error, the process will examine the residue signal for further impulses.

Since the measured channel profiles are relatively static, any multipath component should be present in several consecutive symbols. If a given sample multipath is not detected in several previous or later measured symbols, it can be considered as noise and removed.

One major weakness of this technique is the identification of the first multipath component if the largest multipath is close to the direct component. In Figure 8.7, if the direct component is smaller than approximately 8 dB and before the largest peak, it will be missed. The largest peak would be presented as the direct component, and an error would be introduced into the results.

8.4   NLOS Conditions

8.4.1   Sources of Uncertainty

The main physical implementation details which affect uncertainty regions are:

- oscillator accuracy and drift;
- walls and obstructions increasing apparent path length;
- received signal strength;
- accuracy of delay estimation technique;
- false readings from interference and multipath.

Propagation of the UWB signal through walls will lead to additional delay before arriving at the receiver. Denser wall material will cause the transmitted signal to travel more slowly than the signal would otherwise travel through free space. Transmission through walls will also lead to significant additional attenuation, which will reduce the received strength and lead to increased errors in estimation. The impact of the wall depends on the thickness of the wall and the angle of arrival of the UWB pulse. For very high accuracy positioning systems, this NLOS situation needs to be avoided or compensated for.

The total timing uncertainty at the receiver may be given by

$$\Delta \tau_{total} = \tau_{osc} + \sum_{ii} \tau_i(L, \alpha) + \tau_{meas}(\gamma) \tag{8.19}$$

where τ_{osc} is the timing error introduced by oscillator mismatch; τ_i is the delay introduced by wall ii, and is a function of wall thickness, L and incident angle α, and τ_{meas} is the limit of measurement technique and is a function of signal-to-noise ratio.

The triangulation techniques described above can be enhanced through the use of a number of supporting techniques, including:

- weighting error region based on confidence;
- confidence based on received signal-to-noise ratio and impulse profile;
- log-likelihood measure based on SNR of direct component used for weighting;
- negative log-likelihood measure based on amplitude ratio of direct-to-largest multipath to detect NLOS conditions.

8.4.2 Delay Through Walls

The excess delays caused by the obstructing wall material can be estimated with the help of the environment layout. The propagation τ_a delay is given by the expression

$$\tau_0 = \tau_a + \frac{\sqrt{L^2 + K^2}}{c} = \tau_a + \frac{L}{c \cdot \cos(\alpha)}, \tag{8.20}$$

where τ_0 is the measured delay, L is the antenna separation, K is the distance from normal to wall, c is the velocity of the radio wave and α is the incident angle as shown in Figure 8.8.

A good estimate for the total transmission delay through the wall is

$$\tau = \tau_a + \frac{L_w}{c \cdot \cos(\alpha)} + \tau', \tag{8.21}$$

where τ' is the excess delay caused by the interior wall, L_w is the wall thickness, and α is the incident angle, c is the speed of light in a vacuum.

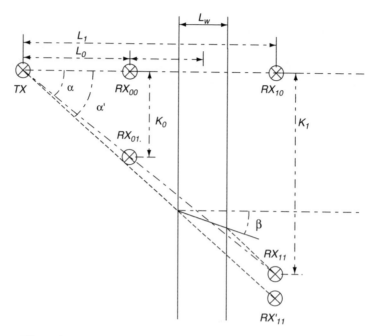

Figure 8.8 Propagation mechanism through isotropic medium

Excess delay τ' can be approximated by an inverse cosine function (from the curve fit):

$$\tau' = \tau_m + \frac{A}{\cos(\alpha)}, \tag{8.22}$$

where τ_m is the delay caused by the medium at incident angle $0°$ and A is an empirical constant. Figures 8.9 and 8.10 show the results of a simple measurement for an indoor environment with a single brick wall (University of Oulu lecture room). The measurements are performed over the range 2 to 8 GHz. Figure 8.9 shows the additional delay compared with free space, while Figure 8.10 shows the additional path loss for the 2–3 GHz, 3–8 GHz and full band versus angle of incidence, where 0 degrees represents the UWB pulse striking the wall at the perpendicular. The additional time delays are of the order of 0.5 to 0.7 ns, which corresponds to 15 to 21 cm of spatial error. The additional path loss values become quite high at higher angles, due to the substantially greater path travelled through the wall.

8.5 Metrics for Positioning

The two most important issues for consideration of the accuracy of a received signal are the signal-to-noise ratio and the number of obstructions the signal has passed through to arrive at the receiver. Obstructions are important as each material affects the speed of light and so increases propagation delay and hence alters the perceived distance from the receiver.

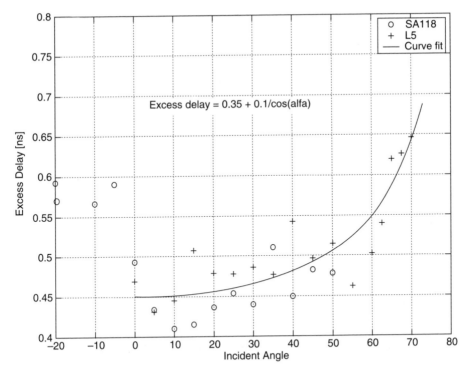

Figure 8.9 Measured additional delay for UWB signal

At the receiver, we are interested in detecting the first received signal component irrespective of the received signal strength. If we have LOS conditions, then the first component will be the strongest and is unaffected by transmission through walls or other obstructions. An example of a LOS channel response profile produced by a ray-tracing simulator is shown in Figure 8.11. In this case, the confidence with which we can detect the correct position is closely related to the signal-to-noise ratio. A high signal-to-noise ratio implies very accurate estimation of the centre of the first received channel path. Let us define a channel confidence metric as

$$P = \log_{10}\left(\frac{|a_0|^2}{\sigma^2}\right) \tag{8.23}$$

where a_0 is the amplitude of the direct signal component and σ^2 is the background noise power.

8.5.1 Identifying NLOS Channels

Later multipath components arrive at the receiver as a consequence of reflection or diffraction caused by objects between the transmitter and the receiver. The additional path travelled and the mechanisms, combined reflection and/or diffraction, introduce

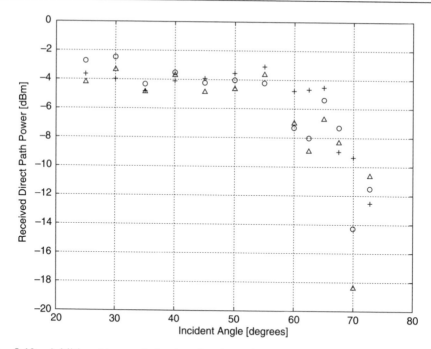

Figure 8.10 Additional transmission loss for through-wall UWB signals (upper △, lower + and total ○ band)

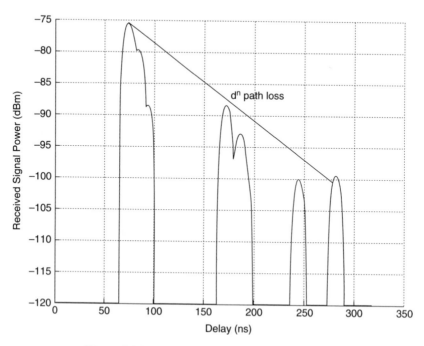

Figure 8.11 Example of LOS multipath profile

additional loss compared with the direct component. Therefore, later multipath components should have lower power than the direct component. Figure 8.12 shows an example of a NLOS channel.

If later multipath components have similar or higher power than the direct component, then the direct component must have travelled through some obstruction, leading to additional attenuation. If a NLOS condition is detected, some additional delay in the propagation of the direct component can be expected, leading to an error in the perceived position. We can include the detection of NLOS conditions in our confidence metric by comparing the amplitude of the direct component to the largest received multipath component.

Given the additional delay of the largest multipath component

$$P = \log_{10}\left(\frac{|a_0|^2}{\sigma^2}\right) + \log_{10}\left(\frac{1}{A_0(\tau_i, \tau_0)}\frac{|a_0|^2}{|a_i|^2}\right) \qquad (8.24)$$

where a_0 is the first signal compnent, a_i is the amplitude of the largest signal component, τ_i is the *total* path delay of the ith signal component, and $A_0(\tau_i, \tau_0)$ is an amplitude scaling factor which accounts for the additional path delay between the direct and the largest signal component. If the direct component is the largest signal component, A_0 is equal to 1.

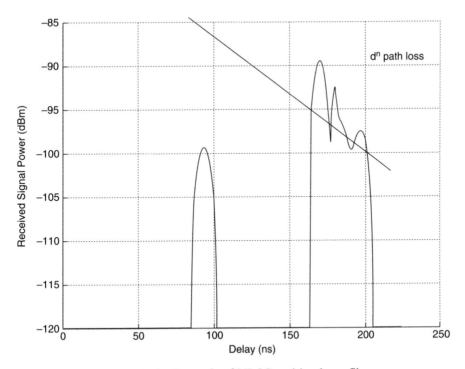

Figure 8.12   Example of NLOS multipath profile

In the event that the received signal is for a NLOS channel, it is possible to estimate the amount by which the direct component has been attenuated and hence the additional delay experienced by the direct component. Let us define A_0 by

$$A_0(\tau_i, \tau_0) = \left(\frac{\tau_i}{\tau_0}\right)^{2n} \tag{8.25}$$

where n is the estimated path loss coefficient. The value of n may be between 2 and 3 for practical systems. Figure 8.13 shows the value of the second term in the confidence metric as a function of the ratio of delays for two equal power channel components.

8.5.1.1 Use of Confidence Metrics

Instead of degrading the confidence of the received signal based on the second term of the confidence metric, the estimate A_0 may be used to estimate the additional attenuation achieved.

In order to detect the position of a user in three dimensions, a minimum of four detectors is required. If many sensors are available, the metrics may be used to select the best detectors to be used for calculating position. Using this approach, the received signal from each sensor can be ranked by confidence. The best four can be used to determine the user position. This approach does assume that at least a crude channel estimate is generated at each sensor.

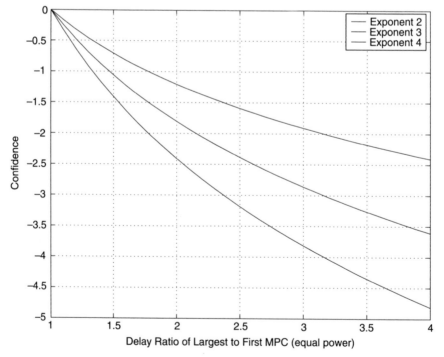

Figure 8.13 Value of second term in confidence metric for equal power channel components

A sensor configuration is illustrated in Figure 8.14, and Figures 8.15 and 8.16 show the confidence weighting for LOS and NLOS conditions for that configuration. The very large uncertainty regions result from the low power, poorly conditioned, impulse response found for the NLOS location.

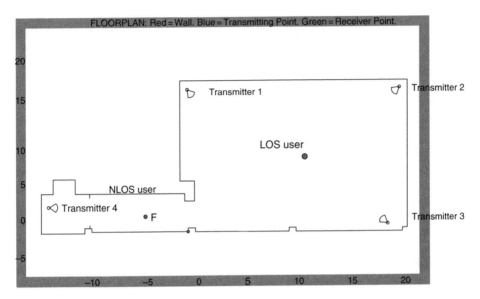

Figure 8.14 Positioning scenario set-up

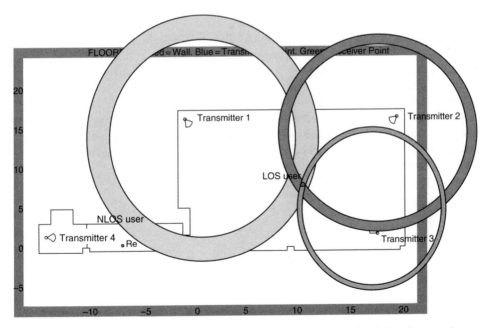

Figure 8.15 Uncertainty regions for LOS user based on received signal strength

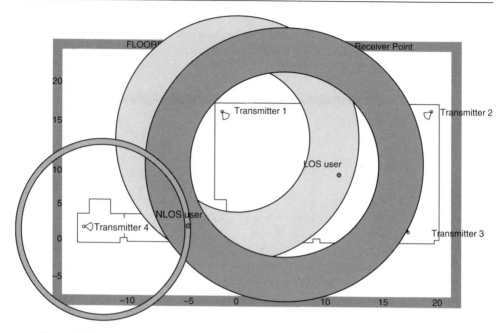

Figure 8.16   Uncertainty regions for NLOS user based on received signal strength

8.6   Conclusions

This chapter has examined some of the most common positioning techniques for radio systems. The wideband nature of the signal allows UWB systems to offer potentially very high positioning accuracy in simple propagation environments. However, the multipath-rich channel and the very low power of the UWB pulse significantly complicate the positioning problem.

Some of the practical problems of UWB positioning have been addressed in this chapter, and some possible means of detecting and possibly avoiding problems with UWB positioning were examined. One very important issue is the identification, and preferably avoidance of, NLOS conditions when performing delay estimation calculations for positioning.

Appendices

Appendix 1

Time hopping pulse position modulated signal in multiple-access case. The formulation is presented in graphic form.

$$s_{tr}^{(k)}(t^{(k)}) = \sum_{j=-\infty}^{\infty} w_{tr}(t^{(k)} - jT_f - c_j^{(k)}T_c - \delta d_{\lfloor j/N_s \rfloor}^{(k)})$$

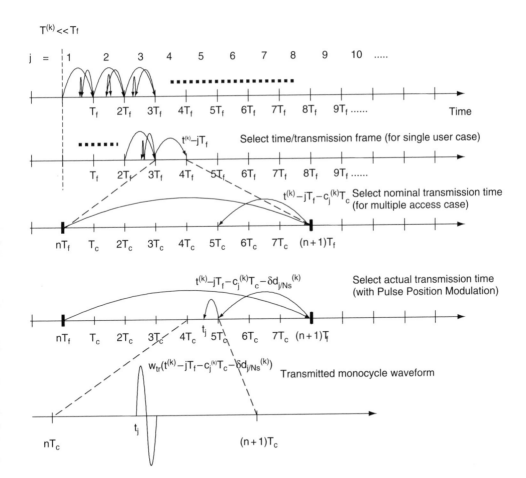

UWB Theory and Applications   Edited by I. Oppermann, M. Hämäläinen and J. Iinatti
© 2004 John Wiley & Sons, Ltd   ISBN: 0-470-86917-8

Meaning of each parameters in the formula.

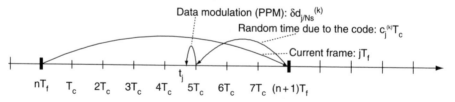

Actual transmission time: $t^{(k)} - jT_f - c_j^{(k)}T_c - \delta d_{j/Ns}^{(k)}$

Data modulation (PPM): $\delta d_{j/Ns}^{(k)}$

Random time due to the code: $c_j^{(k)}T_c$

Current frame: jT_f

nT_f T_c $2T_c$ $3T_c$ $4T_c$ $5T_c$ $6T_c$ $7T_c$ $(n+1)T_f$

t_j

Appendix 2

The frequency mask for indoor UWB devices set by FCC (Federal Communications Commission, 2002b)

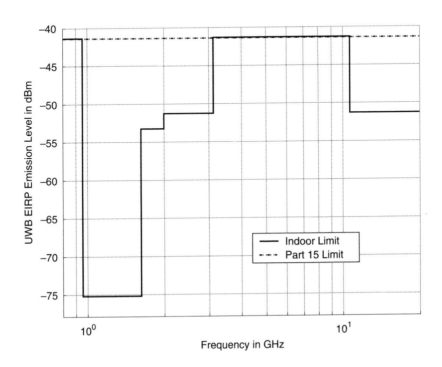

The unit used in the diagram above is dBm/MHz. The next calculations are to convert the unit of signal level from dBm/MHz into power unit dBm/Hz and then into watts (W),

$$-41.3^{\mathrm{dBm}}/_{\mathrm{MHz}} = -101.3^{\mathrm{dBm}}/_{\mathrm{Hz}}$$

$$P_{dBm} = 10\log\frac{P_{watt}}{1\mathrm{mW}}$$

$$\Rightarrow P_{watt} = 10^{\frac{P_{dBm}}{10}} \cdot 1\,\mathrm{mW} = 10^{\frac{-101.3}{10}} \cdot 1\,\mathrm{mW} = 7.413 \cdot 10^{-14}\,\mathrm{W}$$

From this (assuming the gain of antenna is 0 dB) the voltage is calculated in a 50 ohm system ($R = 50\,\Omega$),

$$P = \frac{U^2}{R} \Rightarrow U = \sqrt{PR}$$

$$U_{rms} = \sqrt{7.413 \cdot 10^{-14} \text{ W} \cdot 50 \text{ } \Omega} = 1.925 \text{ } \mu V$$

$$U_p = \sqrt{2} \cdot U_{rms} = 2.72 \text{ } \mu V$$

The result is assumed to be the average voltage level of the transmission. The voltage level of a single pulse is determined from previous by utilising the duty cycle (assumed to be 1%),

$$U_{pulse} = \frac{U_p}{\eta} = \frac{2.72 \text{ } \mu V}{0.01} = 272 \text{ } \mu V$$

Appendix 3

Schematic of UWB transmitter.

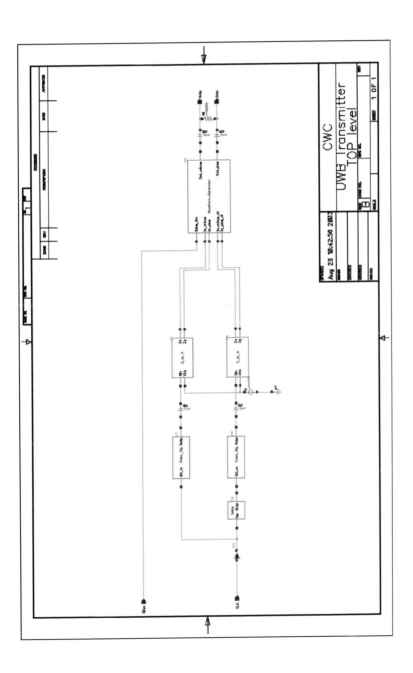

Appendix 4

Schematic of digital pulse generation circuit

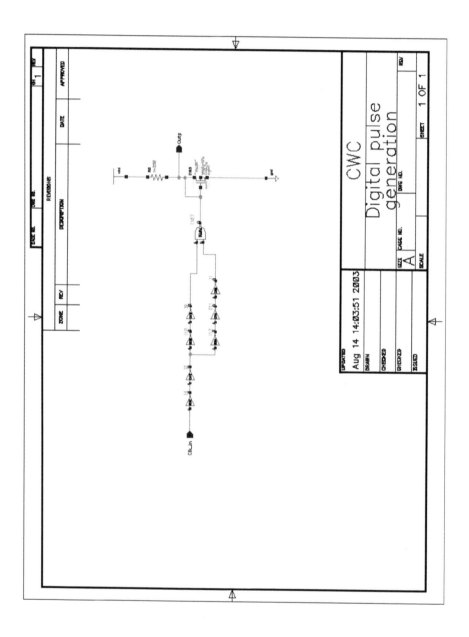

Appendix 5

Schematic of single-ended to differential conversion circuit

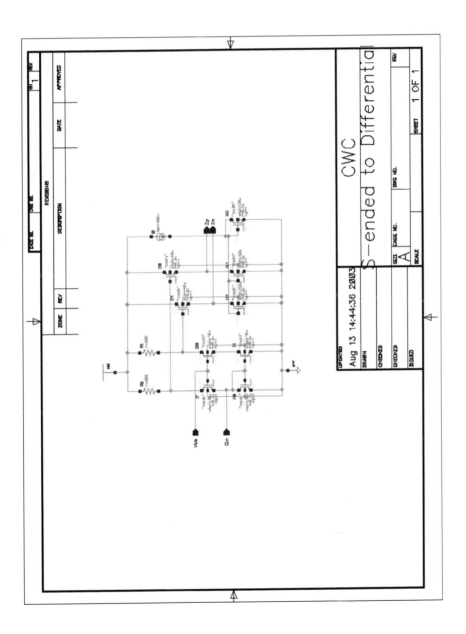

Appendix 6

Schematic of waveform generator

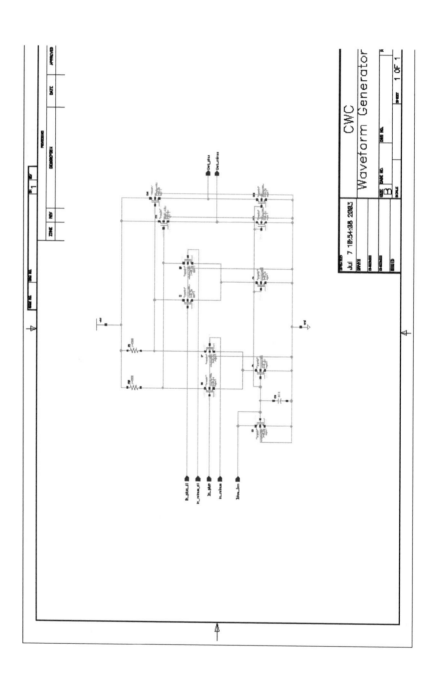

Appendix 7

Schematic of UWB receiver

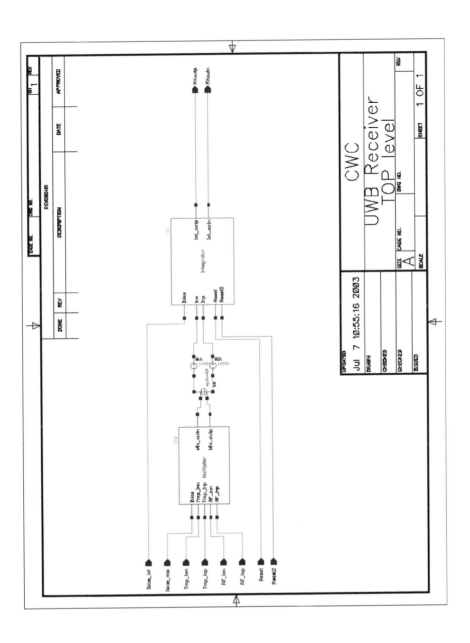

Appendix 8

Schematic of Gilbert multiplier

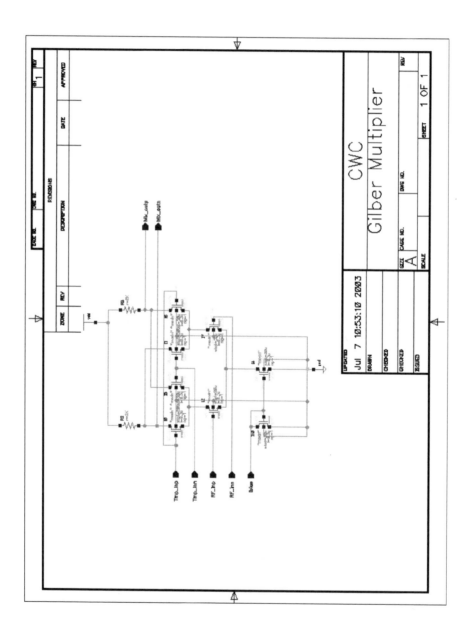

Appendix 9

Schematic of integrator

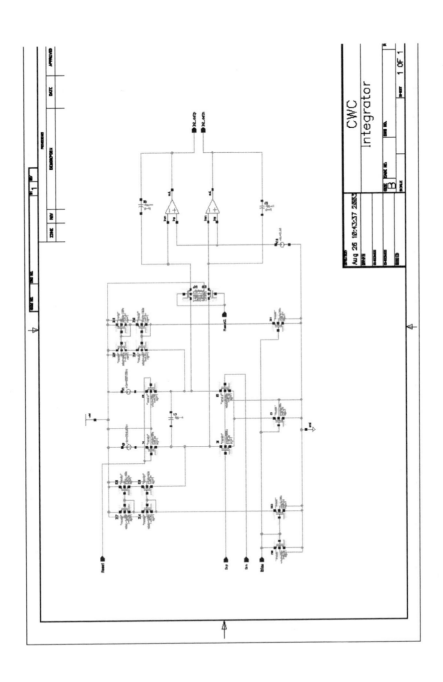

Appendix 10

Schematic of delay element

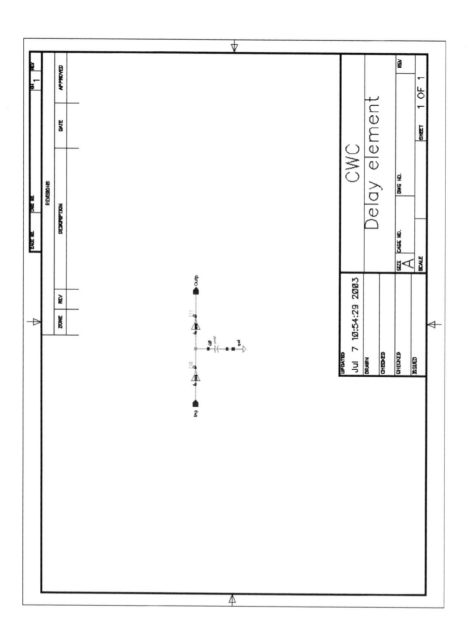

References and Bibliography

Agee, F., C. Baum, W. Prather, J. Lehr, J. O'Loughlin, J. Burger, J. Schoenberg, D. Scholfield, R. Torres, J. Hull and J. Gaudet (1998) 'Ultra-wideband transmitter research', *IEEE Transactions on Plasma Science*, **26**, pp. 860–873.

Allen O. E., Hill D. A., Ondrejka A. R. (1993) 'Time-domain antenna characterisations', *IEEE Transactions on Electromagnetic Compatibility*, **35**, pp. 339–346.

Anderson, F., W. Christensen, L. Fullerton and B. Kortegaard (1991) 'Ultra-wideband beam forming in sparse arrays', *IEE Proceedings H*, **138**, pp. 342–346.

Astanin, L. Y. and A. A. Kostylev (1992) 'Ultra-wideband signals – a new step in radar development', *IEEE AES Systems Magazine*, March, pp. 12–15.

Babanezhad, J. N. and G. C. Temes (1985) 'A 20-V four-quadrant CMOS analog multiplier', *IEEE Journal of Solid State Circuits*, **20**, pp. 1158–1168.

Balanis C. A. (1997) '*Antenna theory: analysis and design*', 2nd edition, John Wiley & Sons, Inc., 941 p.

Baum C. E., Farr E. G. (1993) '*Impulse radiating antennas*', Ultra-Wideband Short Pulse Electromagnetics , H. L. Bertoni, L. Carin, and L. B. Felsen, Eds., New York: Plenum Press, pp. 139–147, 1993.

Baum C. E., Stone A. P. (1993) "*Transient lenses for transmission systems and antennas*", in Ultra-Wideband Short Pulse Electromagnetics , H. L. Bertoni, L. Carin, and L. B. Felsen, Eds., New York: Plenum Press, pp. 211–219, 1993.

Bello, P. A. (1963) 'Characterization of randomly time-variant linear channels', *IEEE Transactions on Communications Systems*, **11**, pp. 360–396.

Bennett, C. and G. Ross (1978) 'Time-domain Electromagnetics and its applications', *Proceedings of IEEE*, **66**, pp. 299–318.

Broyden C. G. (1970) 'The convergence of a class of double-rank minimization algorithms', *Journal of the Institute of mathematics and its applications*, **6**, pp. 76–90.

Buchenauer C. J., Tyo J. S., Schoenberg J. S. H. (1999), '*Aperture efficiencies of Impulse radiating antennas*', in Ultra-Wideband Short Pulse Electromagnetics 4 , E. Heyman, B. Mandelbaum, and J. Shiloh, Eds., Kluwer Academic/Plenum Publishers, New York, pp. 91–108, 1999.

Buchwald, W., A. Balekdjian, J. Conrad, J. Burger, J. Schoenberg, J. Tyo, M. Abdalla, S. Ahern and M. Skipper (1997) 'Fabrication and design issues of bulk photoconductive switches used for ultra-wideband, high-power microwave generation', *The 23rd International Conference on Power Modulators*, Baltimore, MA USA, pp. 970–974.

Burger, J., J. Schoenberg, J. Tyo, M. Abdalla, S. Ahern, M. Skipper and W. Buchwald (1997) 'Development and testing of bulk photoconductive switches used for ultra-wideband, high-power microwave generation', *Proceedings of the 11th IEEE International Pulsed Power Conference*, Baltimore, USA, pp. 965–969.

UWB Theory and Applications   Edited by I. Oppermann, M. Hämäläinen and J. Iinatti
© 2004 John Wiley & Sons, Ltd   ISBN: 0-470-86917-8

Carin, L., N. Geng, M. McClure, J. Sichina and L. Nquyen (1999) 'Ultra-wideband synthetic aperture radar for minefield detection', *IEEE Antennas and Propagation*, **41**, pp. 18–33.

Carin L., Felsen L. (ed.) (1995) *Ultra-wideband, short-pulse electromagnetics 2*. Conference Proceedings. Kluver Academic/Plenium Press, New York, 1995, 605 p.

Cassioli D., Win M. Z., Molisch A. F. (2002) The ultra-wide bandwidth indoor channel: from statistical model to simulations. *IEEE Journal on Selected Areas in Communications*, **20**, Issue: 6, Aug. 2002, Pages: 1247–1257.

Chambers, C., S. R. Cloude and P. D. Smith (1992) 'Wavelet processing of ultra-wideband radar signals', *IEE Colloquium on Antenna and Propagation Problems of Ultra Wideband Radar*, London, UK, 4 p.

Chan Y. T., Ho K. C. (1994) "A simple and efficient estimator for hyperbolic location," *IEEE Transactions on Signal Processing*, **42**, no. 8, pp. 1905–1915.

Chevalier Y., Imbs Y., Beillard B., Andrieu J., Jouvet M., Jecko B., de Legros E. (1999), "*A new broadband resistive wire antenna for ultra wideband applications*", in Ultra-Wideband Short Pulse Electromagnetics 4, E. Heyman, B. Mandelbaum, and J. Shiloh, Eds., Kluwer Academic/Plenum Publishers, New York, pp. 157–164, 1999.

Cohen, M. N. (1991) 'An overview of high range resolution radar techniques', *Proceedings of the National Telesystem Conference (NTC '91)*, Atlanta, GA USA, pp. 107–115.

Conroy, J., J. Locicero and D. Ucci (1999) 'Communication techniques using monopulse waveforms', *Proceedings of the IEEE Military Communications Conference (MILCOM '99)*, Atlantic City, NJ USA, pp. 1181–1185.

Cook, P. G. and W. Bonser (1999) 'Architectural overview of the SPEAKeasy system', *IEEE Journal on Selected Areas in Communications*, **17**, pp. 650–661.

Cramer, R. J.-M., M. Z. Win and R. A. Scholtz (1998) 'Impulse radio multipath characteristics and diversity reception', *Proceedings of the 1998 IEEE International Conference on Communications (ICC '98)*, Atlanta, GA USA, pp. 1650–1654.

Cramer, R. J.-M., M. Z. Win and R. A. Scholtz (1998) 'Evaluation of the multipath characteristics of the impulse radio channel', *Proceedings of Personal, Indoor, and Mobile Radio Communications (PIMRC '98)*, Boston, MA USA, pp. 864–868.

Cramer, R. J.-M., R. A. Scholtz and M. Z. Win (1999) 'On the analysis of UWB communication channels', *Proceedings of the IEEE Military Communication Conference (MILCOM '99)*, Atlantic City, NJ USA, pp. 1191–1195.

Cramer, R. J.-M., R. A. Scholtz and M. Z. Win (1999) 'Spatio-temporal diversity in ultra-wideband radio', *Proceedings of Wireless Communications and Networking Conference (WCNC '99)*, New Orleans, LA USA, pp. 888–892.

Cuomo, K., J. Piou and J. Mayhan (1999) 'Ultra-wideband coherent processing', *IEEE Transactions on Antennas and Propagation*, **47**, 1094–1107.

Daniels J. D. (1996) *Digital design from zero to one*, John Wiley & Sons, Inc. 640 p.

di Benedetto M.-G. (2001) '*Ultra wide band radio system aspects in telecommunications sharing and networking*', Workshop on regulatory issues regarding implementation of Ultra Wide Band Technology in Europe, 20 Mar 2001, Mainz, Germany.

Dickson, D. and P. Jett (1999) 'An application specific integrated circuit implementation of a multiple correlator for UWB radio applications', *Proceedings of the IEEE Military Communications Conference (MILCOM '99)*, Atlantic City, NJ USA, pp. 1207–1210.

di Sorte D., Femminella M., Reali G., Zeisberg S. (2002) 'Network service provisioning in UWB open mobile access networks', *IEEE Journal of Selected Areas in Communications*, **20**, no. 9, pp. 1745–1753

Edgar, D. L., N. I. Cameron, H. McLelland, M. C. Holland, M. R. S. Taylor, I. G. Thayne, C. R. Stanley and S. P. Beaumont (1999) 'Metamorphic GaAs HEMTs with f_T of 200 GHz', *Electronics Letters*, **35** pp. 1114–1115.

Engler, H. F., Jr (1991) 'Systems considerations for large percent-bandwidth radar', *Proceedings of the National Telesystem Conference (NTC '91)*, Atlanta, GA USA, pp. 133–137.

Engler, H. (1993) 'Advanced technologies for ultra-wideband system design', *Proceedings IEEE International Symposium on Electromagnetic Compatibility*, Dallas, TX USA, pp. 250–253.

Excell P. S., Tinniswood, A. D., Clarke R. W. (1998) 'Log-periodic antenna for pulsed radiation', *Electronics Letters*, **34**, pp. 1990–1991.

Fang B. T. (1990) 'Simple solutions for and related position fixes', *IEEE Transactions on Aerospace and Electronic Systems*, **26**, no. 5, pp. 748–753, Sept. 1990.

Farr E. G., Baum C. E., Prather W. D., Bowen L. H. (1999) '*Multifunction impulse radiating antennas: Theory and experiment*', in Ultra-Wideband Short Pulse Electromagnetics 4 , E. Heyman, B. Mandelbaum, and J. Shiloh, Eds., Kluwer Academic/Plenum Publishers, New York, pp. 131–144, 1999.

Federal Communication Commission, FCC (1998) '*Notice of Inquiry*', ET Docket No. 98–153, Rules Regarding Ultra-Wideband Transmission Systems.

Federal Communications Commission (2002a) http://www.fcc.gov/Bureaus/Engineering_Technology/News_Releases/2002/nret0203.html, FCC press release, Feb 2002.

Federal Communications Commission (2002b) '*First Report and Order in the matter of revision of Part 15 of the Commission's rules regarding ultrawideband transmission systems*', ET-Docket 98–153, FCC 02–48, released April 22, 2002.

Fleming, B. (1999) 'Integrated ultra-wideband localizers', *Proceedings of the 1999 International Ultra-wideband Conference*, Washington DC USA, Conference CD.

Fletcher R., Powell M. J. D. (1963) 'A rapidly convergent descent method for minimization', *Computer Journal*, **6**, pp. 163–168, 1963.

Foerster J. (2003) '*Channel modeling sub-committee report – Final*', EEE P802.15 Working Group for Wireless Personal Area Networks (WPANs), Feb 7, 2003.

Foerster J., Green E., Somayazulu S., Leeper D. (2001) 'Ultra-wideband technology for short- or medium-range wireless communications', *Intel Technology Journal*, Q2, 11 p.

Foerster, J., M. Pendergrass and A. F. Molisch (2003) 'A channel model for ultra-wideband indoor communications', *Proceedings of the 6th International Symposium on Wireless Personal Multimedia Communications*, Yokosuka, Japan, pp. 116–120.

Fontana R. J., Larrick J. F., Cade J. E.: '*An ultra wideband communications link for unmanned vehicle applications*', Multispectral Solutions, Inc

Fontana, R. J., Larrick J. F. and J. E. Cade (1997) 'An Ultra-wideband communications link for unmanned vehicle applications', *Association for Unmanned Vehicle Systems International 1997 Conference (AUVSI '97)*, Baltimore, MD USA.

Foster P. R. (1993) '*Reflector antennas for ultra wideband usage*', in Ultra-Wideband Short Pulse Electromagnetics , H. L. Bertoni, L. Carin, and L. B. Felsen, Eds., New York: Plenum Press, pp. 203–209, 1993.

Fowler, C., J. Entzminger and J. Corum (1990) 'Assessment of ultra-wideband technology', *IEEE AES Magazine*, November, pp. 45–49.

Friedlander B. (1987) 'A passive location algorithm and its accuracy analysis', *IEEE Journal of Oceanic Engieering*, **12**, pp. 234–245, Jan. 1987.

Fullerton, L. (1991) 'UWB waveforms and coding for communication and radar', *Proceedings IEEE National Telesystem Conference (NTC '91)*, Atlanta, GA USA, pp. 139–141.

Funk, E., S. Saddow, L. Jasper and A. Lee (1995a) 'Time coherent ultra-wideband pulse generation using photoconductive switching', *Digest of the LEOS Summer Topical Meetings*, pp. 55–56.

Funk, E., S. Saddow, L. Jasper and C. Lee (1995b) 'Coherent power combining of ultra-wideband pulsed radiation in free space', *Microwave Symposium Digest*, Vol. 3, IEEE MTT-S International, Orlando, FL USA, pp. 1299–1302.

Funk, E., S. Ramsay, C. Lee and J. Craven (1998) 'A photoconductive correlation receiver for wireless digital communications', *International Topical Meetings on Microwave Photonics*, Princeton, NJ USA, pp. 21–24.

Gilbert B. (1968) 'A precise four quadrant multiplier with sub nanosecond response', *IEEE Journal of Solid-Stage Circuits*, **3**, pp. 365–373, Dec. 1968.

Gilbert B. (1997) 'A highly linear variant of the Gilbert mixer using a bisymmetric class-AB input stage', *IEEE Journal of Solid-Stage Circuits*, **32**, No. 9, pp. 1412–1223.

Gray P. R., Meyer R. G. (1993) '*Analysis and design of analog integrated circuits*', Third Edition, John Wiley & Sons, Inc. 792 p.

Gregorian R., Temes G. C. (1986) '*Analog CMOS integrated circuits for signal processing*', John Wiley & Sons, Inc. 598 p.

Gill, G. S., H. F. Chiang and J. Hall (1994) 'Waveform synthesis for ultra-wideband radar', *Proceedings of the IEEE Radar Conference*, Atlanta, GA USA, pp. 240–245.

Glisic, S. and B. Vucetic (1997) '*Spread Spectrum CDMA System for Wireless Communications*', Artech House Publisher, Norwood, MA USA, pp. 383.

Griffiths J. (1987) '*Radio Wave Propagation and Antennas: An Introduction*'. Prentice-Hall International, (UK) Ltd., London, UK, 384 s.

Häkkinen J. (2002) '*Integrated RF building blocks for base station applications*', Acta Universitatis Ouluensis, C177, 102 p.

Hämäläinen, M., V. Hovinen and M. Latva-aho (1999a) 'Survey on ultra-wideband systems', COST262/cwc_wg2_td013(99), Thessaloniki, Greece.

Hämäläinen, M., V. Hovinen and M. Latva-aho (1999b) 'Introduction to impulse radio systems', *URSI/IEEE XXIV Convention on Radio Science*, Turku, Finland, 2 p.

Hämäläinen M., Hovinen V., Tesi R., Iinatti J., Latva-aho M. (2002) '*On the UWB system co-existence with GSM900, UMTS/WCDMA and GPS*', IEEE Journal on Selected Areas in Communications, **20**, No. 9, p. 1712–1721.

Hämäläinen M., Hentilä L., Pihlaja J., Nissinaho P. (2003) '*Modified frequency domain radio channel measurement system for ultra wideband studies*', Finnish Wireless Communications Workshop (FWCW 2003), Oulu, Finland.

Harmuth, H. F. (1978) 'Frequency sharing and spread spectrum transmission with large relative bandwidth', *IEEE Transactions on Electromagnetic Compatibility*, Vol. **20**, pp. 232–239.

Harmuth, H. E. and S. Ding-Rong (1993) 'Large current, short length radiator for non-sinusoidal waves', *Proceedings IEEE International Symposium on Electromagnetic Compatibility*, Arlington, VA, USA, pp. 453–456.

Harmuth, H. E. (1984), '*Antennas and Waveguides for Nonsinusoidal Waves*', New York: Academic.

Hovinen, V., M. Hämäläinen and T. Pätsi (2002) 'Ultra-wideband indoor radio channel models: preliminary results', *Proceedings of IEEE Conference on Ultra-wideband Systems and Technologies (UWBST '02)*, Baltimore, MD USA, pp. 75–79.

Hussain, M.(1996) 'An overview of the principles of ultra-wideband impulse radar', *Proceedings of the CIE International Conference on Radar*, Beijing, China, pp. 24–28.

IEEE (2004) http://www.ieee802.org/15/pub/

ITU (2002) '*Preliminary compatibility analysis between space scientific services and UWB*'. ITU Report SE24M16_50 Rev.2

Iinatti, J. (1997) 'Matched filter code acquisition employing a median filter in direct sequence spread-spectrum systems with jamming', *Acta Universitatis Ouluensis*, C102, 54 p.

Iinatti J. (2000) 'Performance of DS code acquisition in static and fading multi-path channel', *IEE Proceedings Communications*, vol.147, Issue 6, December 2000, pp.355–360.

Iinatti, J. and M. Latva-aho (2001) 'A modified CLPDI for code acquisition in multipath channel', *Proceedings of Personal, Indoor, and Mobile Radio Communications (PIMRC'01)*, San Diego, CA USA, pp. F6–F10.

Iverson, D. E. (1994) 'Coherent processing of ultra-wideband radar signals', Radar, sonar, and navigation – *IEEE Proceedings*, **141**, pp. 171–179.

Jakobsson, A., A. Lee Swindlehurst and P. Stoica (1998) 'Subspace-based estimation of time delays and Doppler shift', *IEEE Transactions on Signal Processing*, **46**, 2472–2483.

Johns D. A., Martin K. (1997) '*Analog integrated circuit design*', John Wiley & Sons, Inc. 706 p.

Kardo-Sysoev, A. F., V. I. Brylevsky, Y. S. Lelikov, I. A. Smirnova and S. V. Zazulin (1999) 'Generation and radiation of powerful nanosecond and sub-nanosecond pulses at high pulse repetition rate for UWB systems', 1999 *International Ultra-wideband Conference*, Washington, DC USA, Proceedings on CD.

Katz M. (2002) '*Code Acquisition in advanced CDMA networks*', Acta Universitatis Ouluensis, C175, Finland, 85 p

Keignars, J. and N. Daniele (2003) 'Channel sounding and modelling for indoor UWB communication', *Proceedings of the First International Workshop on Ultra Wideband Systems*, Oulu (IWUWBS '03), Finland, Proceedings on CD, 5 p.

Khorramabadi, H. and P. R. Gray (1984) 'High-frequency CMOS continuous-time filters', *IEEE Journal of Solid State Circuits*, **19**, 939–948.

Kim, A., L. Domenico, R. Youmans, A. Balekdjian, M. Weiner and L. Jasper (1993) 'Monolithic photoconductive ultra-wideband RF device', *Proceedings IEEE International Microwave Symposium*, Atlanta, GA USA, pp. 1221–1224.

Kolenchery, S. S., J. K. Townsend and J. A. Freebersyser (1998) 'A novel impulse radio network for tactical military wireless communications', *Proceedings of the IEEE Military Communications Conference (MILCOM '98)*, Boston, MA USA, pp. 59–65.

Kolenchery, S. S., J. K. Townsend, J. A. Freebersyser and G. Bilbro (1997) 'Performance of local power control in peer-to-peer impulse radio networks with bursty traffic', *Proceedings of the Global Telecommunications Conference IEEE (GLOBECOM '97)*, Phoenix, AZ USA, pp. 910–916.

Kunisch, J. and J. Pamp (2002) 'Measurement results and modelling aspects for the UWB radio channel', *Digest of papers, IEEE Conference on Ultra-wideband Systems and Technologies (UWB ST '02)*, Baltimore, MD USA, 19–23.

König, U. (1999) '*Progress in SiGe heterostructure devices*', *IEE Colloquium on Advances in Semiconductor Devices*, London, UK, pp. 6/1–6/6.

Lai, A. K. Y., A. L. Sinopoli and W. D. Burnside (1992) 'A novel antenna for ultra-wideband applications', *IEEE Transactions on Antennas and Propagation*, **40**, pp. 755–760.

Lamensdorf D., Susman L. (1994) 'Baseband-pulse-antenna techniques', *IEEE Antennas and Propagation Magazine*, **36**, pp. 20–30.

Lang, J. (2003) 'UWB chip design with embedded functionality', UWB Summit, Paris, France, CD Proceedings.

Latva-aho, M. (1998) 'Advanced receivers for CDMA Systems', Acta Universitatis Ouluensis, C125, 179p.

Lee, J. S. and C. Nguyen (2001) 'Novel low-cost ultra-wideband, ultra-short-pulse transmitter with MESFET impulse-shaping circuitry for reduced distortion and improved pulse repetition rate', *IEEE Microwave and Wireless Components Letters*, **11**, pp. 208–210.

Lee J.S., Ngyuen C., Sullicon T. (2001a) 'New uniplanar subnanosecond monocycle pulse generator and transformer for time-domain microwave applications'. *IEEE Transactions on Microwave Theory and Techniques*, **49**, No. 6.

Lee J.S., Nguyen C. (2001b) '*Novel low-cost ultra-wideband, ultra-short-pulse transmitter with MESFET impulse –shaping circuitry for reduced distortion and improved pulse repetition rate*'. *IEEE Microwave and Wireless Components Letters*, **11**, No. 5.

Lestari, A. A., A. G. Yarovoy and L. P. Ligthart (2000) 'Capacitively-tapered bowtie antenna', *Proceedings Millennium Conference on Antennas and Propagation*, Davos, Switzerland, Proceedings on CD.

Lewis, L. R., M. Fasset and J. Hunt (1974) 'A broadband stripline array element', *Digest of the 1974 IEEE Antennas and Propagation Society International Symposium*, pp. 335–337.

Li X. (2002) '*Evaluation of RF CMOS IC technology for wireless LAN applications*', University of Florida, http://www.tec.ufl.edu/~xli. Page available at 25.08.2003.

Lu M., Shi C. (1999) '*Quality ultra-wideband omni-directional antenna*', in Ultra-Wideband Short Pulse Electromagnetics 4, E. Heyman, B. Mandelbaum, and J. Shiloh, Eds., Kluwer Academic/Plenum Publishers, New York, pp. 122–125, 1999.

Lynch, W. C., K. Rahardja and S. Gehring (1999) 'An analysis of noise aggregation from multiple distributed RF emitters', 1999 *International Ultra-wideband Conference*, Washington, DC USA, Proceedings on CD.

Maggio, G., N. Rulkov, M. Sushchik, L. Tsimring, A. Volkovskii, H. Abarbanel, L. Larson and K. Yao (1999) 'Chaotic pulse-position modulation for ultra-wideband communication systems', 1999 *International Ultra-wideband Conference*, Washington, DC USA, Proceedings on CD.

Maloney J. G., Smith G. S. (1993a) 'A study of transient radiation from the Wu-King resistive monopole – FDTD analysis and experimental measurements', *IEEE Transactions on Antennas and Propagation*, **41**, pp. 668–675.

Maloney J. G., Smith G. S. (1993b) 'Optimization of conical antennas for pulse radiation: An efficient design using resistive loading', *IEEE Transactions on Antennas and Propagation*, **41**, pp. 940–947, Jul 1993.

Manabe, T. Takai H. (1992) 'Superresolution of multipath delay profiles measured by PN correlation method', *IEEE Transactions on Antennas and Propagation*, **40**, No 5, 1992, pp. 500–509

McCorkle J.W. (2001) '*Ultra wideband communication system, method, and device with low noise pulse formation*'. World Intellectual Property Organization WO 01/93520 A2.

Moddemeijer R. (1991) 'On the determination of the position of extrema of sampled correlators', *IEEE Transactions on Signal Processing*, **39**, No 1., 1991, pp. 216–291

Montoya T. P, Smith G. S. (1996) 'A study of pulse radiation from several broad-band loaded monopoles', *IEEE Transactions on Antennas and Propagation*, **44**, pp.1172–1182, Aug 1996.

Morgan, M. (1994) 'Ultra-wideband impulse scattering measurements', *IEEE Transactions on Antennas and Propagation*, **42**, 840–846.

Nunnally, N. (1993) 'Generation and application of moderate power, ultra-wideband or impulse signals', *Proceedings International Symposium on Electromagnetic Compatibility*, Dallas, TX USA, pp. 260–264.

Nguyen C., Lee J.– S., Park J. – S. (2001) "Ultra-wideband microstrip quasi-horn antenna", *Electronics Letters*, **37**, pp. 731–732, Jun 2001.

Olhoeft, G. (1999) 'Applications and frustrations in using ground-penetrating radar', 1999 *International Ultra-wideband Conference*, Washington, DC USA, Proceedings on CD.

Parkway P. (2001) '*Ultra-wideband data transmission system*'. World Intellectual Property Organization WO 01/39451 A1.

Petroff, A. (1999) 'Time modulated ultra-wideband: performance on a chip', 1999 *International Ultra-wideband Conference*, Washington, DC USA, Proceedings on CD.

Petroff, A. and P. Withington (2000) 'Time modulated ultra-wideband: overview', 1999 *International Ultra-wideband Conference*, Washington, DC USA, Proceedings on CD.

Pochain G. P. (1999) '*Large current radiator for the short electromagnetic pulses radiation*', in Ultra-Wideband Short Pulse Electromagnetics 4 , E. Heyman, B. Mandelbaum, and J. Shiloh, Eds., Kluwer Academic/Plenum Publishers, New York, pp. 149–155, 1999.

Polydoros, A. (1982) 'On the synchronization aspects of direct-sequence spread spectrum systems', Ph.D. dissertation, University of Southern California, Los Angeles, California, USA.

Prasad, R. and S. Hara (1996) 'An overview of multi-carrier CDMA', *Proceedings of the Fourth IEEE Symposium on Spread Spectrum Techniques and Applications (IS SSTA 96)*, Mainz, Germany, 107–114.

Proakis J. G. (1995) '*Digital communications*', McGraw-Hill Inc., Singapore, 928 s.

Proakis J. G., Salehi M. (1994) '*Communication system engineering*', Prentice-Hall, Inc.

Qiu, R. (1998) 'A theoretical study of the ultra-wideband wireless propagation channel based on the scattering centres', *Proceedings of the IEEE Conference on Vehicular Technology (VTC 98)*, Ottawa, ON Canada, pp. 308–312.

Raines, J. (1999) 'Cumulative impact of TM-UWB devices and effects on three aviation receivers', 1999 *International Ultra-wideband Conference*, Washington, DC USA, Proceedings on CD.

Ramirez-Mireles, F. and R. A. Scholtz (1997) 'Performance of equicorrelated ultra-wideband pulse-position-modulated signals in indoor wireless impulse radio channels', in *Proceedings of the Sixth IEEE Pacific Rim Conference on Communications, Computers and Signals*, Victoria, BC Canada, pp. 640–644.

Ramirez-Mireles, F. and R. A. Scholtz (1998a) 'System performance analysis of impulse radio modulation', *Proceedings of the Radio and Wireless Conference*, Colorado Springs, AZ USA, pp. 67–70.

Ramirez-Mireles, F. and R. A. Scholtz (1998b) 'Multiple access with time hopping and block waveform PPM modulation', *Proceedings of the 1998 IEEE International Conference on Communications (ICC '98)*, Atlanta, GA USA, pp. 775–779.

Ramirez-Mireles, F. and R. A. Scholtz (1998c) 'Multiple-access performance limits with time hopping and pulse position modulation', *Proceedings of the IEEE Military Communications Conference(MILCOM '98)*, Boston, MA USA, pp. 529–533.

Ramirez-Mireles, F. and R. A. Scholtz (1998d) 'Wireless multiple access using SS time-hopping and block waveform pulse position modulation, Part 1: signal design', *Proceedings of the International Symposium on Information Theory and Its Applications* (ISITA), Mexico.

Ramirez-Mireles, F. and R. A. Scholtz (1998e) 'Wireless multiple access using SS time-hopping and block waveform pulse position modulation, Part 2: system performance', *Proceedings of the International Symposium on Information Theory and Its Applications* (ISITA), Mexico.

Ramirez-Mireles, F., M. Z. Win and R. A. Scholtz (1997a) 'Signal selection for the indoor wireless impulse radio channel', *Proceedings of the IEEE Conference on Vehicular Technology (VTC 97)*, Phoenix, AZ USA, pp. 2243–2247.

Ramirez-Mireles, F., M. Z. Win and R. A. Scholtz (1997b) 'Performance of ultra-wideband time-shift-modulated signals in the indoor wireless impulse radio channel', *Proceedings of the Thirty-first Asilomar Conference*, Pasific Grove, CA USA, pp. 192–196.

Roberts, R. "XtremeSpectrum CFP Document", IEEE P802.15, IEEE P802.15–03/154r3, 124 p., Jul 2003.

Rothwell, E., K. Chen, D. Nyquist and J. Ross (1995) 'Time-domain imaging of airborne targets using ultra-wideband or short-pulse radar', *IEEE Transactions on Antennas and Propagation*, **43**, 327–329.

Rowe, D., B. Pollack, J. Pulver, W. Chon, P. Jett, L. Fullerton and L. Larson (1999) 'A Si/Ge HBT timing generator IC for high-bandwidth impulse radio applications', *Proceedings of the IEEE Custom Integrated Circuits Conference*, San Diego, CA USA, pp. 221–224.

Schau H. C., Robinson A. Z. (1987) 'Passive source localization employing intersecting spherical surfaces from time-of-arrival differences', *IEEE Transactions on Acoustics, Speech, and Signal Processing*, **35**, pp. 1223–1225, Aug. 1987

Scholtz, R. A. (1993) 'Multiple access with time-hopping impulse modulation', *Proceedings of Military Communications Conference (MILCOM '93)*, Boston, MA USA, pp. 447–450.

Scholz, R. A. 2nd M.Z. Win (1997) 'Impulse radio', in 'Wireless communications – TDMA versus CDMA', S. Glisic and P. A. Leppänen (eds.), Kluwer Academic Publishers, London, pp. 245–263.

Scholtz, R. A., R. Cramer and M. Win (1998) 'Evaluation of the propagation characteristics of ultra-wideband communication channels', *Proceedings of the IEEE Antennas and Propagation Symposium*, Atlanta, GA USA, pp. 626–630.

Shlager K. L., Smith G. S., Maloney J. G. (1994) 'Optimization of bow-tie antennas for pulse radiation', *IEEE Transactions on Antennas and Propagation*, **42**, pp. 975–982.

Shlager K. L., Smith G. S., Maloney J. G. (1996) 'Accurate analysis of TEM horn antennas for pulse radiation', *IEEE Transactions on Electromagnetic Compatibility*, **38**, pp. 414–423.

Shlivinski, E. Heyman, and R. Kastner (1993) 'Antenna characterization in the time domain', *IEEE Transactions on Antennas and Propagation*, **45**, pp.1140–1149.

Silva, J. A. N. da, and M. L. R. de Campos (2002) 'Orthogonal pulse shape modulation for impulse radio', *Proceedings of the International Telecommunications Symposium (ITS 2002)*, Brazil.

Smith J. O., Abel J. S. (1987) 'Closed-form least-squares source location estimation from range difference measurements', *IEEE Transactions on Acoustics, Speech, and Signal Processing*, **35**, pp. 1661–1669.

Staderini, E. (1999) 'Medical applications of UWB radars', 1999 *International Ultra-wideband Conference*, Washington, DC USA, Proceedings on CD.

Stickley, G. F., D. A. Noon, M. Chernlakov and I. D. Longstaff (1997) 'Preliminary field results of an ultra-wideband (10–620 MHz) stepped-frequency ground penetrating radar', *Geosciences and Remote Sensing (IGARSS '97)*, Singapore, pp. 1282–1284.

Talvitie J. (1997) 'Wideband radio channel measurement, characterisation and modelling for wireless local loop applicatios', Acta Universitatis Ouluensis Technica C99, 94 p.

Taylor J. D. (ed.) (1995) '*Introduction to ultra wideband radar systems*'. CRC Press, Inc., Boca Raton, FL USA, 670 p.

Theron, I., E. Walton, S. Gunawan and L. Cai (1999) 'Ultra-wideband noise radar in the VHF/UHF band', *IEEE Transactions on Antennas and Propagation*, **47**, pp. 1080–1084.

Time Domain Corporation (1998) '*Comments of time domain corporation**, Docket 98–154. In the Matter of Revision of Part 15 of the FCC's Rules Regarding Ultra wideband Transmission Systems.

Time Domain Corporation (2003) '*PulsOn technology overview*', http://www.timedomain.com/. Page available at 25.08.2003.

Tiuraniemi S. (2002) '*Method and arrangement for generating cyclic pulses*'. United States Patent Application 10/304915.

Torrieri D. J. (1998) 'Statistical theory of passive location systems', *IEEE Transactions on Aerospace and Electronic Systems*, 20, no. 2, pp. 183–198.

Ultra Wideband Working Group (1998) '*Comments of the ultra wideband working group, Docket 98–153*'. In the Matter of Revision of Part 15 of the FCC's Rules Regarding Ultra wideband Transmission Systems.

Ultra Wideband Working Group (1999) '*The 1999 International Ultra Wideband Conference Proceedings-CD*', Huntsville, AL USA.

Ultra Wideband Working Group (2004) http://www.uwb.org/regulatory/regulatory.html.

van Cappellen W. A., de Jongh R. V., Ligthart L. P. (2000) 'Potentials of ultra-short-pulse time-domain scattering measurements', *IEEE Antennas and Propagation Magazine*, **42**, pp. 35–45, Aug 2000.

Vaskelainen, L. and R. Pitkäaho (2002) Antennas Transmitting or Receiving Broadband Impulses', VTT Information Technology, Espoo, Finland, Internal report.

Vickers, R. (1999) 'Design and applications of airborne radars in the VHF/UHF band', 1999 *International Ultra-wideband Conference*, Washington, DC USA, Proceedings on CD.

Weeks, G. and J. Townsend (1999) 'Quantifying the covertness of impulse radio', 1999 *International Ultra-wideband Conference*, Washington, DC USA, Proceedings on CD.

Weeks, G., J. Townsend and J. Freebersyser (1999) 'Performance of hard decision detection for impulse radio', *Proceedings of the IEEE Military Communications Conference (MILCOM '99)*, Atlanta, GA USA, pp. 1201–1206.

Weissoerger, M. (1982) *An Initial Critical Summary of Models for Predicting the Attenuation of Radio Waves by Trees*, ITT Research Institute, Report number: A343811, Annapolis, MD USA, 162 p.

Wicks M. C., Antonik P. (1993) '*Polarization diverse ultra-wideband antenna technology*', in Ultra-Wideband Short Pulse Electromagnetics , H. L. Bertoni, L. Carin, and L. B. Felsen, Eds., New York: Plenum Press, pp. 177–187, 1993.

Win M.Z., Scholtz R. A. (2000) 'Ultra-Wide Bandwidth Time-Hopping Spread-Spectrum Impulse Radio for Wireless Multiple-Access Communications', *IEEE Transactions on Communications*, **48**, pp. 679–689.

Win, M. Z. (1999) 'Spectral density of random time-hopping spread-spectrum UWB signals with uniform timing jitter', *Proceedings of the IEEE Military Communications Conference (MILCOM '99)*, Atlanta, GA USA, pp. 1196–1200.

Win, M. Z. and Z. Kostic (1999) 'Impact of spreading bandwidth on rake reception in dense multipath channels', *Proceedings of the 1999 IEEE International Conference on Communications (ICC '99)*, Vancouver, BC Canada, pp. 78–82.

Win, M. Z. and R. A. Scholtz (1997a) 'Comparisons of analogue and digital impulse radio for wireless multiple access communications', *Proceedings of the 1997 IEEE International Conference on Communications (ICC '97)*, Montreal, Canada, pp. 91–95.

Win, M. Z. and R. A. Scholtz (1997b) 'Energy capture vs correlator resources in ultra-wideband width indoor wireless communications channels', *Proceedings of IEEE Military Communications Conference*, Monterey, CA USA, pp. 1277–1281.

Win, M. Z. and R. A. Scholtz (1998a) 'Impulse radio: how it works', *IEEE Communication Letters*, **2**, pp. 36–38.

Win, M. Z. and R. A. Scholtz (1998b) 'On the energy capture of ultra-wideband width signals in dense multipath environments', *IEEE Communications Letters*, **2**, pp. 245–247.

Win, M. Z. and R. A. Scholtz (1998c) 'On the robustness of ultra-wideband width signals in dense multipath environments', *IEEE Communications Letters*, **2**, pp. 51–53.

Win, M. Z. and J. H. Winters (1999a) 'On maximal ration combining in correlated Nakagami channels with unequal fading parameters and SNRs among branches: an analytical framework', *Proceedings of the Wireless Communications and Networking Conference*, New Orleans, LA USA, pp. 1058–1064.

Win, M. Z. and J. H. Winters (1999b) 'Analysis of hybrid selection/maximal ratio combining of diversity branches with unequal SNR in Rayleigh fading', *Proceedings of the IEEE Conference on Vehicular Technology (VTC '99)*, Houston, TX USA, pp. 215–220.

Win, M. Z. and J. H. Winters (1999c) 'Analysis of hybrid selection/maximal ration combining in Rayleigh fading, *Proceedings of the 1999 IEEE International Conference on Communications (ICC '99)*, Vancouver, BC Canada, pp. 6–10.

Win, M. Z., R. A. Scholtz and L. W. Fullerton (1996) 'Time-hopping SSMA techniques for impulse radio with an analogue modulated data sub-carrier', *Proceedings of the IEEE Symposium on Spread Spectrum Techniques and Applications (ISSSTA '96)*, Mainz, Germany, pp. 359–364.

Win, M. Z., F. Ramirez-Mireles, R. A. Scholtz and M. A, Barnes (1997a) 'Ultra-wideband width signal propagation for outdoor wireless communications', *Proceedings of the IEEE Conference on Vehicular Technology (VTC '99)*, Phoenix, AZ USA, pp. 251–255.

Win, M. Z., J.-H. Ju, V. O. K. Li and R. A. Scholtz (1997b) 'ATM-based ultra-wideband width multiple-access radio network for multimedia PCS', *Proceedings of IEEE Fourth Annual Networld + Interop Conference*, Las Vegas, Nevada USA, pp. 101–108.

Win, M. Z., R. A. Scholtz and M. A, Barnes (1997c) Ultra-wideband width signal propagation for indoor wireless communications', *Proceedings IEEE International Conference on Communications (ICC '97)*, Montreal, Canada, pp. 56–60.

Win, M. Z., G. Chrisikos and N. Sillenberger (1999) 'Impact of spreading bandwidth and diversity order on the error probability performance of rake reception in dense multipath channels', *Proceedings of Wireless Communications and Networking Conference*, New Orleans, LA USA, pp. 1558–1562.

Withington P. (2004) '*Impulse Radio Overview*'. http://www.time-domain.com.

Withington, P., R. Reinhardt and R. Stanley (1999) 'Preliminary results from ultra-wideband (impulse) scanning receivers', *Proceedings of the IEEE Military Communications Conference (MILCOM '99)*, Atlantic City, NJ USA, pp. 1186–1190.

Yarovoy, A. G., R de Jongh and L. Ligthart (2000) 'Ultra-wideband sensor for electromagnetic field measurements in time domain', *Electronics Letters*, **36**, 1679–80.

Yngvesson, K. S., T. L. Korzniowski, Y.-S. Kim, E. L. Kollberg and J. F. Johansson (1989) 'Tapered slot antenna – a new integrated element for millimetre wave applications', *IEEE Transactions on Microwave Theory and Techniques*, **37**, 365–74.

Ziolkowski, R. W. (1992) 'Properties of electromagnetic beams generated by ultra-wideband width pulse-driven arrays', *IEEE Transactions on Antennas and Propagation*, **40**, pp. 888–905.

Index

UWB Theory and Applications Edited by I. Oppermann, M. Hämäläinen and J. Iinatti
© 2004 John Wiley & Sons, Ltd ISBN: 0-470-86917-8